GALILEO, WAKE UP!
They've gone mad...

Beyond Space-Time: Dialogues for a New Physics

Olivier SERRET

GALILEO, WAKE UP!

ISBN: 979-8-3221728-2-6

Legal deposit: February 2024

Self-publisher: Olivier SERRET

47330 Cahuzac

o.serret@free.fr

Printed on demand by Amazon

Copyright © Olivier Serret, 2024

All rights reserved - No part of this publication may be reproduced.

PREFACE

> "It's *about examining the value of the arguments on both sides without deciding anything, then deferring to the judgment of those who know more than we do.*"

Galileo

In ancient Greece, a shepherd came to the high priest of the Temple and asked:
"We're taught that when the thunder rumbles, it's the god who's angry, and when it rains, it's his wife, the goddess, who weeps. Is it true about these stories?"
"What else do you want it to be?" replied the priest.
"I don't know."
"See, that's what we're telling you!"

Today, it seems absurd that gods can make rain and shine. The thunderbolt is understood to be a flow of electrons, even if we don't know exactly what the elusive electron looks like - we don't even know its speed or position at the same time. And perhaps tomorrow a new explanation will complete it, or even replace it.

Above all, we remain helpless in the face of what we don't understand. **We need an explanation for** the physical phenomena in our environment. Like this shepherd, we may have questions about physical phenomena close to the speed of light. But the official theory, that of Relativity, advocated by the self-proclaimed guardians of the temple of Physics, cannot be questioned by ordinary mortals. As absurd as it may seem, we have to believe that the passage of time depends on the speed of each individual, and that the lengths of objects also depend on the speed of each individual! Even if the experiments and observations tending to prove this complex theory are still tenuous and subject to complex interpretations.

<u>The aim of this book</u> is to present in an offbeat way and synthesize some twenty articles published in online scientific journals dealing with an alternative solution, Neo-Newtonian Mechanics; for scientific readers, the references of these articles are given in Footnotes.

This book is necessary in the sense that one can only constructively criticize the official theory if one has an alternative solution to propose. For the reader or listener who might finally be prepared to accept criticism of relativistic evidence invariably asks the following question: *What do you propose instead?* Without this sesame, criticism of relativistic evidence alone proves, in practice, unacceptable. That's why this book proposes an alternative solution, bearing in mind that there are dozens, if not hundreds, of others.

This work is also sufficient in the sense that, for the alternative to be acceptable, the official explanation must also be deconstructed. For, as Galileo puts it through the mouth of the conservative-minded Simplicio, "*First demonstrate to me that these appearances cannot be accounted* *for by supposing the Sun mobile and the Earth stable, otherwise I shall not abandon my belief in the movement of the Sun and the immobility of the Earth*"[1] .

My previous book[2] offered a serious, critical and well-sourced analysis of relativistic evidence. In fact, this was the fault that was returned to me with that first book: too serious in form, a sort of mini-thesis; and chapters, analyzing the various experiments and observations, independent of each other. This is why the form of the present book is somewhat different. First of all, it's fiction. Its narrative, spread over four days, is inspired by Galileo's Dialogue. It's an ongoing dialogue which, while it may resemble a play in form, is also a book about physics and cosmology. And while there are no treasures to be conquered as in an adventure book, you will find a progression through the chapters, from our earthly experiences to observations from the far reaches of the Universe.

Basically, this book presents neo-Newtonian mechanics, which, unlike other alternative theories, is in line with Newtonian mechanics. There are no extravagant assumptions. Today, 99% of physics problems are solved by Newtonian mechanics, so there's no need to appeal to the theory of Relativity. Of course, the proponents of this theory of Relativity will say that, at non-relativistic speeds, the results of their theory are the same as those of Newtonian mechanics. The results, yes, but not the foundations! In our everyday environment, according to the theory of Relativity, the person travelling by train would age less quickly than the person remaining on the platform, and an apple would fall to the ground through deformation of space-time! Whereas with neo-Newtonian mechanics, the basic concepts remain the same as those of Newtonian mechanics: the passage of time is independent of the observer's speed, and the fall of bodies depends on the forces of gravity.

---

[1] Galileo, *Dialogue sur les deux grands systèmes du monde*, Ed. du Seuil - 1992, page 522.
[2] O. Serret, *Would Einstein still be right? 10 experiments for or against Relativity*, Ed. Amazon - 2022.

In short, the structure of the dialogue allows us to compare the relativistic and neo-Newtonian representations of the world. This representation is called neo-Newtonian because, compared **with classical Newtonian mechanics, only one point changes. You then have to unwind the string to discover all its consequences.** That's what I invite you to do on the following pages.

PRELIMINARIES (GALILEAN)

LAW, PRINCIPLE, REFERENCE FRAME AND GALILEAN TRANSFORMATION

0. THE MAIN PROTAGONISTS

Galileo Galilei, *born in Italy in 1564, was a mathematician, physicist and astronomer. He is famous for having developed the telescope and for his observations that confirmed Copernicus' heliocentric model, thus challenging the geocentric model of the Universe. His ideas were controversial, and he was condemned by the Catholic Church as a heresy for his support of heliocentrism. Despite this, his pioneering work in mechanics laid the foundations of modern physics.*

Johannes Kepler, *born in Germany in 1571, was an astronomer and mathematician renowned for his laws of planetary motion. He developed a method based on the principle of elliptical trajectories. His work laid the foundations of modern astronomy and contributed to the abandonment of the geocentric model of the Universe.*

Isaac Newton, *born in England in 1642, was a physicist, mathematician and astronomer. He is famous for having formulated the laws of motion and the law of universal gravitation, thus laying the foundations of classical physics. His work had an immense impact on many areas of the physical sciences, and revolutionized our understanding of the solar system.*

Albert Einstein, *born in Germany in 1879, was a theoretical physicist. He is best known for his theories of special and general relativity, and above all for his famous equation $E=mc^2$. His work revolutionized modern physics and our understanding of the cosmos.*

Noé *is a fictional character, in charge of presenting and defending neo-Newtonian mechanics. He adopts the author's positions.*

The President *is a fictional character whose role is to guide the debates.*

1. FALLING BODIES

Arcetri (near Florence), 1642.

GALILEO: Eppur si muove! (And yet it moves!). No, I didn't dare say it, or even whisper it, after the sentence of the Tribunal of the Sacred Congregation of the Holy Office. Were they really reproaching me for trying to understand how Heaven works without providing any convincing arguments? Or that, as a simple layman despite my years of novitiate, I had dared to interpret a biblical verse[3], like a Reformed with the transubstantiation of the host? Or rather, to have disobeyed the injunction of the Cardinal of the Inquisition and the directive of the Pope, who felt deceived? Or, quite simply, to have simplistically and ironically reported, in their opinion, the papal position[4]? Never mind, damn those Dominican dogs! Some forty years ago, they burned Giordano Bruno alive. In addition to their grievances against his support for the Copernican conviction and his speculations on other inhabited worlds, which diminished the importance of Man in divine Creation, they surely feared his questioning of the divinity of Christ as well as his theory of the reincarnation of souls, but they burned him alive anyway because he refused to recant! After twenty days, including three days of stormy interrogation, in order to save my skin, or simply to avoid being subjected to the questioning, it was better to give up in the end and keep a low profile. Did I even think about it at the time? Not really, or yes of course. It's my innermost conviction, forged over the years... Ouch,

---

[3] Joshua 10:13. "So the sun stood still, and the moon stopped, till the nation avenged itself on its enemies, as it is written in the Book of Jashar. The sun stopped in the middle of the sky and delayed going down about a full day."

[4] "A very learned and eminent person, and one who needs to be calmed down."

this pain won't let me go. These stab wounds in the shoulder, it's terrible. And these rheumatisms that have been nagging me for decades, hampering my movements. It's my body that's aging, not my mind. But putting up with the pain takes up my mind and what little energy I have left. For two months now, I haven't had the strength to leave this bed. And I haven't been able to see for a long time now, perhaps five years since I last saw the sun's rays. The last straw for a member of the Lynx Academy! Even as a youngster, my eyes were a pain. My long observations with the telescope probably tired them out. And my discovery of the rotation of sunspots, and therefore of the Sun, probably didn't help either, even if, with a few exceptions, I only observed their projection. Ah, time flies! Seventy-seven years already, I've barely seen them go by. Come on, I've had an interesting life. Despite relationship difficulties with my mother and a strict religious upbringing, I had a carefree youth until the death of my father. A varied education in Florence, the most beautiful city in the world, steeped in the arts and Platonic philosophy. Then Padua, just a few miles from Venice, where freedom of expression and religion reigned supreme. I worked there as a simple mathematics teacher and conducted various experiments. It was the place of my best eighteen years. It was there that I met a Venetian woman of low status, but so charming, who was to become the mother of my children. When I returned to Florence, I had to leave her in Padua, where she died some time later. She gave me beautiful children. Two pretty girls who understood nothing of the world. One, a sweet girl, is now dead, may she rest in peace. The other, half hysterical, doesn't speak to me much any more; she's still angry at me for putting her in a convent. And yet, also an illegitimate daughter, she was hardly marriageable. And a son who finally bears my name and, after years of conflict, is beginning to take an interest in my work with the pendulum and the measurement of time. Not forgetting my various companions who have brought me moments of happiness and gentleness. And, finally, my various babies: my experiments and my books on the workings of the world. One major experiment concerned the fall of the deep. But who's coming? Ah, it's you, Torricelli and Viviani. Your loyalty to me comforts me.

TORICELLI: How do you feel, illustrious master?

GALILEO: Oh, illustrious master, you say. You will surpass me, my disciples, or so I hope. Perhaps you'll discover other things about Nature. No, I'm sure of it. Right now, I'm feeling weaker and weaker. I was remembering the fall of the bass, but words don't come as easily. Can you tell me again what you remember?

TORICELLI: Of course! Idiots and pedants think that the heavier it is, the faster it falls. In their defense, even the great Aristotle wrote that heavier objects fall faster than lighter ones. You've dared to question the teachings of this ancient genius. **You have shown, by dropping balls of cork and lead or ebony, that they fall to the ground at the same speed. Provided, of course, that the air friction is not too great, as it can be with a feather. That's why you take objects of the same shape, balls. In this way, the fall of heavy bodies in a vacuum is independent of their mass.**

VIVIANI: That experience was from the top of the Leaning Tower of Pisa, in front of the masters of the University of Pisa, wasn't it?

GALILEO: No, it's a lovely legend, and one that I confess I helped to propagate. In fact, it was only from the bell tower of a church in Padua, with friends and family. Above all, I measured the fall of bodies using an inclined plane of my own making. I showed and measured it. I was also able to measure that the speed of fall was not uniform, contrary to popular belief, but increased with time. **I also demonstrated that all bodies fall in the same way.** Let's take several cork balls, holding hands as it were, whose total weight would be identical to that of the ebony ball. As they are together as heavy as the ebony ball, this set of cork balls falls, at the same speed as the ebony ball. So each cork ball falls at the same speed as the ebony ball. All bodies fall at the same speed. But there's no substitute for experience to validate this reasoning. I'm tired of talking and need to rest. I'm afraid I must ask you to leave, and thank you again for your visit.

They go out.

GALILEO: Where was I? My experiences, my achievements, my observations, my books. To my half-surprise, the five hundred copies of my book "*Sidereal Messenger*" sold out in a matter of days. A turning point in my life. At the age of 45, I went from engineer and experimenter to astronomical discoverer and founder of an entirely new science of Natural Philosophy. The development of the telescope is undoubtedly my most resounding achievement, with the help of the Murano glass masters. In any case, it was this lunette that started me on the road to financial ease when I sold my rights to the Serenissima Republic of Venice. What a bargain! It doubled my emoluments. And it was above all the Grand Duke of Tuscany who allowed me to return to my native region and freed me from the obligation to teach. I was thus able to devote myself to observing the stars, Jupiter and its satellites of course, Venus and its occultations, and the Moon. Ah, this Moon, which, contrary to popular belief, is far from crystalline, with its cavities and protuberances. It has mountains higher than those on Earth! How similar it is to our Earth. I'm even convinced it's inhabited. But how do we get in touch with these Lunians? I'm getting tired, so let's try to get some sleep and regain some strength.

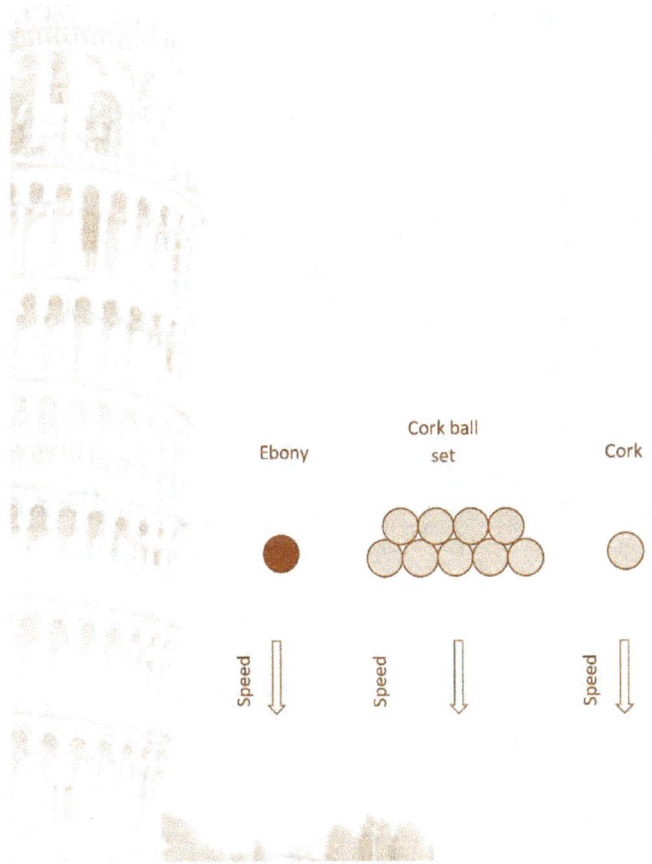

FIGURE 1: FALLING BODIES

GALILEO, WAKE UP!

2. THE GALILEAN PRINCIPLE AND TRANSFORMATION

THE VISITOR: Greetings, O illustrious celestial messenger. We come from another land, another time. Fear not, we have come as friends. Your fame is great. Your principle will live on.

GALILEO: Hello, mage from elsewhere. You say my principle will survive me, but which one? I've enunciated various precepts. Are you talking about the fact that the Sun is at the center of the world?

THE VISITOR: No, but the principle used with your boat example. The boat moves at a constant speed relative to the water. You imagined butterflies enclosed in a cabin. It's impossible for them to know whether the boat is moving or not.

GALILLE: Yes, that's right. I was saying that uniform motion is like nothing, i.e. that the state of uniform motion is, for a shipboard observer, indistinguishable from the state of rest.

THE VISITOR: **We'll call this 'Galileo's principle' and its statement will be: the motion of an object in translation, in other words in a straight line, remains unchanged if an additional parallel uniform motion is added to it.**

GALILEO: My boat example seems to me to be expressed more simply. Straight-line motion is only an approximation on a small scale. That's why I was talking about uniform circular motion, like that of the stars.

THE VISITOR: The consequence of your principle is that, if a person aboard the boat is also walking in the same direction as the boat's motion with a constant speed relative to the boat, then relative to an outside observer on the shore, the person would appear to be moving with **a speed that is the mathematical sum of the two motions, that of the boat and that of the person. We'll call this Galileo's speed transformation.**

GALILEO: Indeed, mathematics helps us to understand the world. Is this why you've come to visit me?

THE VISITOR: Actually, that's not the main reason. The Inquisition's trials of you will make you a martyr, especially in secular times. The procedures used were unjust. You were threatened with torture and perhaps death, and you had to recant. It was only a debate of ideas, and you almost paid for it with your life. The object of the trial was to decide between two worlds: the Earth at the center of the world or the Sun at the center of the world. Now, in the world I come from, we find ourselves in the same situation, debating between two visions. One called Relativity, the other Neo-Newtonian Mechanics.

GALILEO: I don't know either of them, and their very names seem strange to me, especially the second one. How can I help you?

THE VISITOR: For two reasons. On the one hand, because both claim to follow your principle, Galileo's principle. And secondly, because who better to ensure the fairness of a debate than someone who has been unjustly accused?

GALILEO: Thank you for your concern, but no, I'm too old to undertake any travel at my age.

THE VISITOR: We understand perfectly well. In fact, it's more like your spirit is traveling with us.

GALILEO: How could it be possible? And my mind itself is quite tired, it has lost its lucidity, its liveliness.

THE VISITOR: Of course, we also have the means to invigorate your spirit.

GALILEO: It sounds like magic, or worse, witchcraft. Taking this leap into the unknown doesn't seem to me to be in line with the Holy Scriptures, and it's going to get me into even more trouble. Talk to me instead about the fundamental idea that I hold so dear and that almost cost me my life: that of Copernicus, with the Sun at the center of the world. Will my arguments be accepted?

THE VISITOR: Yes and no, actually. Of course, it will be recognized that the Earth rotates on itself, and your argument about the trade winds will be taken up. Your observation of Jupiter's moons shows that stars can revolve around a body other than the Earth. By observing the phases of Venus, you've also shown that it orbits the Sun.

GALILEO: I don't understand, what are you saying? That, despite my arguments, it will be recognized that not everything revolves around the Earth, that the Earth turns on itself, but that it would nevertheless remain at the center of the world? Or more precisely, according to Ptolemy's system, around a fictitious point, the center of the deferent? In this way, the Sun would revolve around the Earth or this point, and the planets around the Sun or this point. So it would be this nonsense of a Tycho-Brahe model that would finally be retained?

THE VISITOR: You know what follows from your principle. Speed doesn't exist in itself; motion is merely conventional in relation to an observer. It's the same thing to say that the Earth is moving in relation to the Sun as to say that it's the Sun that's moving in relation to the Earth. Both can be true at the same time, depending on where the observer is located. This is why your principle is also known as the principle of relativity.

GALILEO: I agree when there are only two objects. But let me remind you that there are many stars. The mathematically simplest frame of reference turns out to be the one where the trajectory of the majority of stars follows a perfect circle around the Sun, not the Earth.

THE VISITOR: The Tycho-Brahe model is, relative to a terrestrial observer, indeed more complex but, from this point of view, it's just as correct. However, it's the model with the planets, including the Earth, orbiting the Sun that will prevail.

GALILEO: So the Copernican model I defend will be favored. So why did you say that my arguments won't be fully accepted?

THE VISITOR: For example, if the Earth does indeed turn on itself, its double rotation, daily and annual, would only really explain one tide a day, even with the backwash effect. However, on the ocean, tides generally occur twice a day, at variable times depending on the Moon, and with varying intensity. On the other hand, planetary trajectories are not circular with epicycles, but rather elliptical.

GALILEO: So, Tycho-Brahe would be right about the tides and his successor Kepler about the elliptical trajectories of the stars. And, in a way, the Extraordinary Commission of the Tribunal of the Holy Office, which rejected my arguments, wouldn't have been entirely wrong. It just goes to show that I can't help you.

THE VISITOR: On the contrary, the search for Truth in the reading of the things of Nature is subtle, and many of your arguments are judicious. Within the Church, you'll have a mausoleum in a church, and even a charismatic pope will call you a "genius physicist", recognizing that there can be no opposition between Faith and Science. But what will indisputably separate the two models, between Copernicus' and Tycho-Brahe's, is the aberration of the stars. It's the Earth that moves against the background of the stars, not the Sun, or very little of it.

GALILEO: So, if in accordance with my principle, which you call relativity, all these frames of reference are equivalent for describing the laws of Nature, there is still one that is to be preferred. In the case of the course of the stars, Copernicus' is the simplest.

THE VISITOR: Your ability to analyze and demonstrate, your keen sense of observation and your unfortunate experience of an unjust conviction can make you an excellent arbitrator in this field.

GALILEO: And you were saying that your two competing theories are both based on this other principle, that of boats... I had expressed this in response to those who said that the Earth couldn't move, because then objects would fall vertically. Fools! I wonder how my principle will be used. Still, it excites my curiosity and my desire to debate it, as I always have. I already feel younger. Well, so be it! Let's go, I'll accompany you to your ship.

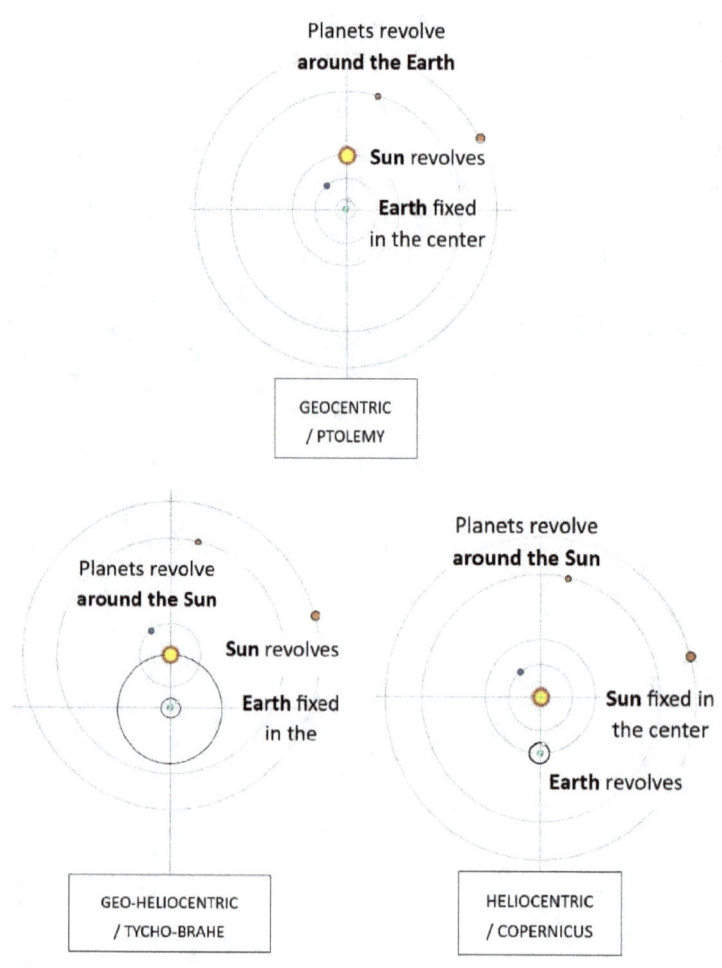

FIGURES 2: PTOLEMY, TYCHO-BRAHE AND COPERNICUS REFERENCE FRAMES

3. THE LAW OF INERTIA AND GALILEAN REFERENCE FRAMES

GALILEO, *re-entering the ship, to himself*: The cockpit of this ship appears quite vast, the lighting is subdued and of changing colors, there's a smell of camphor, it's a good omen. Ah, that's it, I feel we're leaving.

THE VISITOR: I'm delighted you've chosen to accompany us. Do you know that one of your laws will outlive you, that of inertia?

GALILEO: What do you mean, another of my precepts?

THE VISITOR: Yes, that other one, the law of inertia: **In the absence of external influence, every body persists in uniform rectilinear motion.**

GALILEO: That must be it, even if I expressed it differently. As I said earlier, rectilinear motion is only a small-scale approximation to circular motion. This horizontal rectilinear motion, once imparted to a mobile, is indelibly imprinted with speed. To move forward at a constant speed, you don't need to make a constant effort; the initial movement is enough. This may seem counter-intuitive. If there were no air or ground resistance, this speed would remain constant. To demonstrate this, I used the example of a galley. If a stone is dropped from the top of a mast, whatever the speed of the galley, the stone falls at the foot of the mast, not behind it. This is because the stone has acquired the speed of the galley.

THE VISITOR: This law of inertia is a consequence of your principle. A movement exists only in relation to an observer, to a frame of reference. **And the frames of reference where this law of inertia applies are called Galilean or inertial frames of reference.**

GALILEO: In these frames of reference, we can't tell if we're moving, we only perceive changes in speed.

THE VISITOR: In fact, that's what's happening on this ship. We're traveling at a constant speed, and we're not even aware of it.

GALILEO: That's right. *To himself:* But what's going on? We seem to be stopping. Have we arrived yet? Oh no, here we go again. Look, the door's opening. Who's that puny fellow with the goatee coming in? Who could it be?

KEPLER: Hello, I'm Johannes Kepler, imperial astrologer and mathematician. I dreamed of boarding a sailing caravel powered by an ether wind to the Moon. Unless, of course, it was to Jupiter. But I think I recognize Professor Galileo. At last, I meet you, Professor Galileo Galilei.

GALILEO: Good morning, Mr Kepler. Apart from passing on a few anagrams, I deliberately did not reply to your last letter. While I appreciated your support in your letter "*Conversation with the Astral Messenger*", I nevertheless found your speculations there to be fanciful and dangerous. Moreover, engaging in an epistolary exchange with a German reformer would have entailed risks for me personally. Already eighteen years before my second trial, I had to deny to the Roman Holy Office a letter, sent to a mathematician friend, in which I criticized the Church's dogmatic position on Copernicanism.

KEPLER: I've known you to be less cautious. The distribution of one of your previous books to the general public, who hardly understood it, together with your essentially polemical attitude, had already resulted in the first trial in which Copernicus' work, including his almanac of unparalleled accuracy, was judged to be misleading and contrary to Scripture, and therefore put on the index.

GALILEO: Come on, it wasn't the first time the Church had strongly condemned theses it deemed subversive, as were those of Siger of Brabant, who wanted to dissociate knowledge from faith. He paid for this with his life.

THE VISITOR: And unfortunately, in the future, there will be more censorship of books for religious, political or moral reasons.

GALILEO: At the second trial, my disobedience in not promoting Copernicanism was, in my opinion, no more than a pretext. I think what I was really accused of was showing that the supralunar world to which souls ascend was not perfect and divine, as Giordano Bruno had already assumed. My description of Nature was independent of the revelations of the Holy Scriptures, and so, for them, my description was absurd, false and, above all, heretical, probably the worst of crimes for the Church. In the background, there was also resentment at my earlier comments on atomism, which could have undermined the transubstantiation of the host. The Catholic Church, guardian of Faith and Knowledge, had already suffered enough from the spread of the Reformation in the Holy Roman Empire, and wanted to crack down in its own lands to reinforce its authority.

KEPLER: Germanic lands weren't much friendlier to religious freedom than Latin ones. Remember the fate that John Calvin reserved for Michel Servet. And Martin Luther called Copernicus mad. I also had to move three times and give up my wife's property to avoid recanting my Protestant faith in the face of the Counter-Reformation. I also had to fight for over a year in a witchcraft trial against my mother, who was literally born under a bad star. It could have happened to me directly, as one of my aunts was also burned at the stake.

GALILEO: You have indeed shown great courage on several occasions. However, let's also recognize that the Church wasn't always hostile towards us, and we both benefited from the support of Jesuit scholars on occasion. It could also be that I've misjudged you, or rather that I was wrong to call your hypotheses on the tides childish. For I've just been told that the waters of the sea are actually attracted to the Moon at a distance. So the Moon would have some kind of occult magnetic action on the water, although I'm still skeptical.

THE VISITOR: Allow me to intervene, gentlemen. The reason is a little different, it's called the law of universal attraction. The Moon attracts the Earth with its oceans, which swell under its effect, creating the tide.

GALILEO: Yet the waters of the Mediterranean, except at its eastern extremities like Venice, do not rise with the course of the Moon. You mentioned that there are generally two tides a day. The attraction of the Moon may explain the variable tide times, but that too only explains one tide a day.

THE VISITOR: You're right, O Galileo. I note that your mind remains sharp. It's actually a combination of two phenomena. The first, as we've just said, is the effect of the Moon's attraction. The ocean is deformable, and the part facing the Moon is more attracted to it, hence the first bulge. The second bulge is on the opposite side of the Earth. The explanation would be quite similar to the one you had, with the Earth rotating on itself in addition to its orbit, which would generate a centrifugal force resulting in a local increase in speed and hence this second bulge.

GALILEO: So, according to you, my explanation wasn't entirely wrong, although I do understand that it wasn't entirely correct. However, in your centrifugal explanation, would this second bulge have the same amplitude as the first bulge? If these two effects are generated by two different causes, there's no reason why they should have the same value, is there?

THE VISITOR: I'm interrupting our discussion at just the right moment, because we've stopped again to take on board our ship the discoverer of this universal law of attraction I've just mentioned. The door opens and there he is.

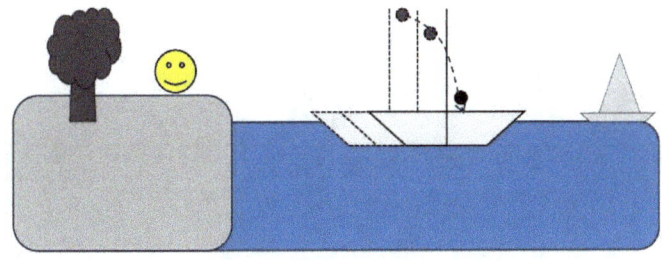

Reference frame (of the observer) = the ground.
Seen from the ground, parabolic free fall

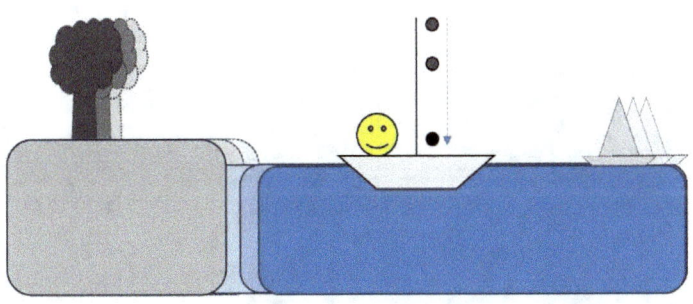

Reference frame (of the observer) = the ship.
Seen from the ship, straight free fall

FIGURES 3: GALILEAN REFERENCE FRAMES

GALILEO, WAKE UP!

4. INERTIAL MASS AND GRAVITATIONAL MASS

NEWTON: Hello, let me introduce myself, Sir Newton, born in 1642 and knighted at the age of 62 for the extraordinary quality of my work and discoveries, particularly in celestial mechanics, which I was able to put into equations.

THE VISITOR: Good morning Sir Newton. You're just in time, we were just discussing universal attraction and more specifically the phenomenon of tides.

GALILEO: Good morning, sir. This principle of an attraction at a distance seems occult to me. How can the Moon act at a distance on the oceans?

NEWTON: This also remains one of my big questions. How can gravitational attraction be exerted at a distance? I'm still looking for the answer in alchemical treatises...

GALILEO: In alchemical treatises? For me, alchemy is just magic.

NEWTON: With all due respect to a physics giant such as yourself, alchemy, contrary to popular belief, is not simply the quest for the philosopher's stone that transmutes lead into gold or confers immortality. The ancient treatises of alchemy must surely conceal secrets handed down from the original wisdom, which we need to know how to decipher. Our voyage in this vessel may well enable us to learn more about immortality and time in general.

GALILEO: The point I was trying to make about the uselessness of alchemy to explain attraction at a distance is that fragments of matter should be able to move through the ether and influence each other by direct contact. Ether is enough, no need for alchemy, don't you think?

NEWTON: But alchemy also deals with substances, essences, souls and virtues. The behavior of the smallest bodies could complete my planetary system. The answer probably lies in the infinitely small, those fragments of matter as you call them.

GALILEO: And what's your main discovery?

NEWTON: According to my law of universal attraction, all bodies attract each other in proportion to their respective masses.

GALILEO: In proportion to their respective masses, you say. However, you only have to observe the free fall of an ebony ball and a cork ball to realize that they fall at the same speed. It's not the heavier one that falls faster, nor is the fall in proportion to their respective masses, contrary to what many believe.

NEWTON: It's **that we have to consider both aspects of the phenomenon, the gravitational mass and the inert mass. On the one hand, gravitational mass is the mass that attracts and is attracted. This is my law of universal gravitation. On the other hand, there's inertial mass, which opposes the variation of motion.**

GALILEO: What are you getting at?

NEWTON: **To this. An ebony ball is more attracted to the Earth than a cork ball, because its gravitational mass is greater. Gravitational attraction is in proportion to gravitational mass. But, at the same time, since the inertial mass of the lead ball is greater than that of the cork ball, it requires more effort to set it in motion. And as you pointed out, with the exception of air friction, both balls fall at the same speed. This means that if they are of different natures, the serious mass and the inert mass are exactly equal in value. This mathematical equality is clearly demonstrated by the march of the stars.**

GALILEO: Getting back to our subject, how do you relate these two masses to the tides?

NEWTON: The tide is the bulge of the ocean at both ends of the globe. The first bulge, facing the Moon, is easily explained by the gravitational pull of the Earth's solid mass, which is stronger than at the center of gravity. The second bulge, on the opposite side, the so-called antipodal bulge, is explained by the gravitational attraction being weaker than at the Earth's gravitational center, which might seem paradoxical.

GALILEO: Isn't it possible that this is due to a centrifugal force on the Earth's periphery?

NEWTON: Anything can be imagined, but I'll end my explanation by showing you the link between tides and masses. For example, on the antipodal side, the gravitational force is less than the inertial force. The resultant pulls in the opposite direction to the Moon. The opposite is true on the side facing the Moon.

GALILEO: Well, I'm a bit confused, but I'll think about it.

Galileo takes a back seat.

GALILEO: It's disturbing, because the explanation of the tides as an effect of the Earth's motion is the keystone of my demonstration to prove that the Sun is immobile at the center of the world. I devoted the fourth and final day of my Dialogue to this. If the tides are caused by the Moon, then all my arguments are over, and the Earth could be at the center of the world, according to Tycho-Brahe's model. Listening to him, this Mr Newton has found other answers to all the phenomena of the Universe. And he makes fun of me by calling me a physics giant, even if it's not entirely untrue, it's still sycophancy. In short, he's not very nice to me. He doesn't seem to have all his wits about him either, presenting himself as having been born in the year we are, 1642. The wigged man's attire is bizarre, a little too sophisticated. If he's the basis of neo-Newtonian mechanics, I'm afraid I have a rather unfavorable opinion of this theory.

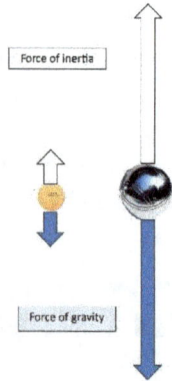

FIGURE 4: FORCE OF GRAVITY & FORCE OF INERTIA

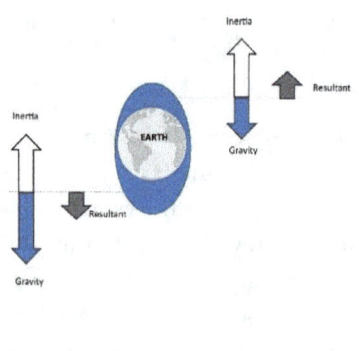

FIGURE 5: TIDES

5. THE CONSTANT SPEED OF LIGHT

GALILEO: One more stop, we'll never reach the Moon! Who's coming now? Ah, he seems a nice chap, with his tousled head and big moustache. On the other hand, his outfit is most bizarre - simple, but very bizarre. Let's hear him introduce himself.

EINSTEIN: Hello, my name is Albert. It's a great honor to be introduced to the experts of the Physics Evaluation Group. A very special greeting to you, Sir Isaac Newton. In practice, my theory doesn't contradict yours, it just extends it to speeds close to those of light. But when we are at normal speeds, my theory precisely recovers the predictions of your theory.

NEWTON: Thank you.

EINSTEIN: And I salute you, Galileo Galilei. Did you know that I have taken up your two principles of relativity and inertia? Your principle of relativity of motion is one of the two postulates that underpin the first part of my theory of Relativity, the one restricted to so-called Galilean reference points. And I also used your work on falling bodies to postulate mass equivalence in the second, more general, part of my theory of Relativity.

GALILEO: Good morning, Mr Albert. I'm delighted. But to come back to your original point, I had tried, in vain, to measure the speed of light with lanterns positioned less than a mile apart. If it's not infinite, it must be extremely great. How great would this speed of light be, please?

EINSTEIN: Well, knowing that the speed of a cannonball is about 300 meters per second, the speed of light is a million times that speed.

GALILEO: That's impressive. I wonder how anyone could measure such a speed. Physical phenomena at this speed must be extremely rare, it makes you wonder on what occasion you use it. But that's not my other question. I'm flattered to hear you say that you've taken my so-called Galileo principle as a postulate, modesty aside. Nevertheless, let's call it the principle of relativity. Could you explain the second postulate of the first part of your theory of Relativity, the one restricted to Galilean reference points?

EINSTEIN: With great pleasure. **The second postulate of Special Relativity is the principle of the invariance of the speed of light in any inertial or Galilean reference frame.**

GALILEO: This seems to me to follow from my principle of relativity and its associated mathematical rule. The speed of a cannonball is 300 m/s, whether fired from the ground or from a moving boat. As I told my students, if the cannonball is fired from the ground, its velocity is 300 m/s relative to the ground. If the ball is fired from a moving boat, its speed is 300 m/s relative to the boat. And in this second case, if fired towards the bow of the boat, which is moving at 1 m/s relative to the shore, as seen from the shore, the speed of the cannonball is the mathematical sum, i.e. 301 m/s.

EINSTEIN: No, the second postulate shows that the behavior of light is different from that of a cannonball. This didn't exist in your time, but let's take self-moving cars. Self-propelled means that the car moves by itself, without being pulled by a horse. The same reasoning would apply to a horse-drawn carriage, but the automobile allows higher speeds, so this will be more demonstrative. Take, for example, a car travelling at 30 m/s.

GALILEO: You've got a great imagination, young man, but I'll let you get on with it.

EINSTEIN: When a second car, for example travelling at 35 m/s, overtakes the first car travelling at 30 m/s, the occupants of the first car have the impression that the second car is catching up with them at a speed of 5 m/s.

GALILEO: This is simply the consequence of my principle of relativity: the person would seem to be moving with a speed that is the sum of the two movements, that of the boat and that of the person.

EINSTEIN: Indeed! Now for another case. When a third car arrives at a speed of 35 m/s, but in the opposite direction to the first car which is travelling at 30 m/s, the occupants of the first car have the impression that the third car is coming towards them at a speed of 65 m/s.

GALILEO: That's always the consequence of my principle of relativity.

EINSTEIN: Absolutely. This is the originality of my second principle, that of the invariance of the speed of light. Light behaves differently from physical objects. Light is made up of photons, tiny grains of light. They are emitted at a speed of 300,000,000 m/s, which we'll call c speed.

GALILEO: I understood that 300,000,000 m/s is their speed relative to the ground.

EINSTEIN: **That's what we'll see. Now, let's take a car driving away from a candle or a light source. The occupants of the car will see the light from this light source coming towards them at the speed of 300,000,000 m/s".**

GALILEO: **You're wrong, the photon will arrive at a speed of 299,999,970 m/s.**

EINSTEIN: **No, no, that's what my second postulate means. In the same way, the occupants of a car going in the direction of the same light source will see the photon arrive at them at the speed of 300,000,000 m/s, not 300,000,030 m/s".**

GALILEO: By what magic? Whatever the direction of the car, its occupants would always see the photon emitted by the light source arriving at the same speed relative to them, whether they were not moving or moving very fast. This is extremely strange. It even contradicts the addition of velocities in my principle of relativity, doesn't it?

EINSTEIN: No, because the photon has no mass, it's a different principle. And it also applies to all massive bodies approaching the speed limit of light. Because nothing can go faster than this speed of 300,000,000 m/s, whatever our frame of reference.

These words plunged Galileo into a sea of perplexity.

GALILEO: This otherwise charming man seems to be crazy. Unless I am. Newton says he was born in the year we are, he talks about self-driving cars moving at high speeds. In fact, I now have the feeling that we're not going to the Moon, but that it's my mind that's traveling into the future. What will I discover there? Can I ever go back to my disciples? They'll never believe me. But the debates promise to be interesting, so let's wait and see.

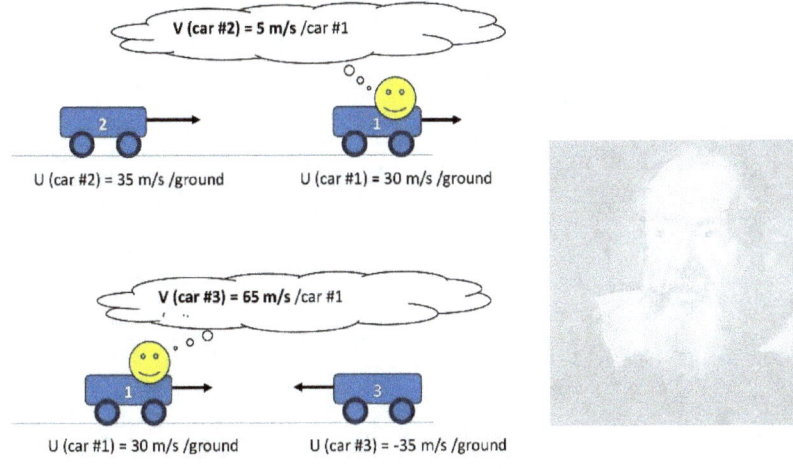

FIGURE 6: GALILEO'S PRINCIPLE OF RELATIVITY

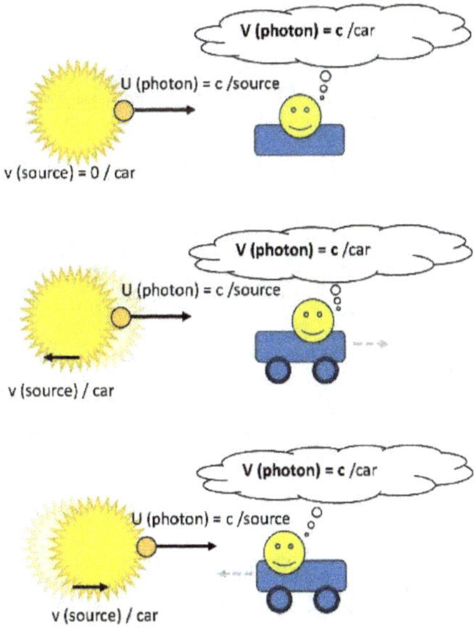

FIGURES 7: SECOND POSTULATE OF RELATIVITY

GALILEO, WAKE UP!

DAY ONE (RELATIVITY)

FUNDAMENTAL RELATIVISTIC EXPERIMENTS

GALILEO, WAKE UP!

6. THE THEORY OF SPECIAL AND GENERAL RELATIVITY

The atmosphere is solemn. An imposing portrait of Einstein marks the opening of the first day of IPPC, the Intergenerational Panel on Physics Change.

THE PRESIDENT: Good morning, ladies and gentlemen. I declare open this Commission of the IPPC, the Intergenerational Panel on Physics Change, which I have the honor of chairing. We will be bringing in experts from different generations, different eras and different countries, chosen for their skills and also for the development of their own experiences, in order to refine the representation, we have of the world of physics. Today, the theory of Relativity prevails unchallenged, and has long been able to provide us with explanations and predictions for the physical phenomena we encounter. But science is constantly progressing, and knowledge is expanding. The physical world is what it is, and any theory is no more than an approximation to it. A theory is valid only as long as another, more precise, theory does not evolve or even replace it. In the case of physics, this was, in very simplified terms, Aristotle's theory, which originated in Antiquity and prevailed throughout the Middle Ages, i.e. for almost two millennia. It was replaced by Newtonian mechanics, which provided comprehensive and satisfactory answers for two centuries, accompanying the industrial revolution. This was in turn replaced by the current theory of Relativity, which opens the doors to the cosmos. It is in this spirit that we are going to examine a new theory that aims to succeed it. Or, to be more precise, we won't be examining the theory in detail, as we don't want to enter into a debate of ideas. As far as possible, we'll confine ourselves to facts, experiments and observations, which will obviously have to be interpreted in the light of each of these two theories. Over the next few days, I'm going to hear from a number of experts who have carried out remarkable experiments and observations in these fields of physics and cosmology. I'll be accompanied by a number of undisputed personalities. While the great philosopher Aristotle was unable to join us, we are delighted to welcome Professor Galileo, experimenter and, it could be said, founder of physics as a modern science. Professor Kepler, an immense mathematician who pragmatically extracted the three great laws governing planets and satellites from astronomical surveys. Sir Newton, an outstanding mathematician, who put the world into equations. And, of course, Professor Einstein, who revolutionized our perception of the world and is honoring us with his presence. And a little unknown, Noé, an independent researcher, who will explain his theory, neo-Newtonian mechanics, and who will be exceptionally authorized to take part in these hearings. But before we begin, I'd like to give you a brief introduction to each of the two theories whose evidence we'll be examining. Without further ado, I give you the floor, dear Professor Einstein.

EINSTEIN: Thank you, Madam Chairman, and thank you, dear colleagues. I won't take up too much of your time, and I'll keep my remarks brief. It was at the age of 26 that I was able to have my founding article on what we call Special Relativity published in a leading journal. It's restricted to inertial reference frames, still called Galilean in honor of Professor Galileo, whom we have the pleasure of welcoming among us.

GALILEO: The honor is mine, Sirs.

EINSTEIN: Please. In fact, its first postulate is Galileo's principle of relativity, with which you are all familiar, and which I have reformulated as follows: the laws of physics have the same form in all Galilean reference frames. My theory differs from the results of Newtonian mechanics at very high speeds, known as relativistic speeds, close to the speed of light. To achieve this, I had to introduce a second postulate, that of the constancy of the speed of light in any inertial or Galilean reference frame. The direct consequence is that time and distance are no longer constant, but variable, depending on the observer.

GALILEO: Just a question. Throughout my life, I've fought against Aristotle's theory, which separated the terrestrial and celestial realms in order to unify our vision of the world, whether terrestrial or planetary. By relying on two apparently antagonistic principles, one for low speeds, the other for very high speeds, aren't you tending to split the uniqueness of our worldview?

EINSTEIN: Your question is pertinent. This second postulate was added to complete and back up your first postulate. In fact, some experiments - we'll talk about them again, I imagine - called into question your transformation of speeds when using light. That's why we had to supplement your postulate with this one, bearing in mind that light has the characteristic of having no mass.

THE PRESIDENT: Thank you for that clarification, please continue.

EINSTEIN: At the time, this theory was restricted to uniform rectilinear systems, and was not intended to be general. What's more, it was inappropriate for describing falling bodies. It therefore took me ten years of research to complete it, culminating in General Relativity. Gravitation is now seen as an inertial system, devoid of the forces of attraction at a distance.

NEWTON: So you've solved the problem I've been trying in vain to find the answer to in alchemical treatises - the problem of the force at a distance, which doesn't exist after all.

EINSTEIN: Indeed, the general explanation is the curvature of space-time by matter.

THE PRESIDENT: Thank you for that introduction to Relativity, which we appreciated for its synthesis and brevity, as requested. Let's now hear from our new fellow student.

GALILEO, *aside:* Relativity is the flow of variable time, it's still very special.

GALILEO, WAKE UP!

7. NEO-NEWTONIAN MECHANICS

NOÉ: Thank you, Madam Chairman, for welcoming me to your esteemed assembly for these days, in the presence of all these eminent scientists, to examine the possible merits of this alternative solution. As you said, every theory has its own weaknesses and needs to be supplemented, or even replaced, by another theory.

THE PRESIDENT: That's right, but I don't want to point out the rare shortcomings of Relativity - we'll come back to that when we hear from the experts. It's just a question of briefly explaining what makes your theory so original.

NOÉ: Well. This theory is called neo-Newtonian mechanics, because it's right in line with Newtonian mechanics.

NEWTON: I'm happy with it, but what makes it different?

NOÉ: On one particular point, the equality between serious mass and inert mass, which doesn't apply.

NEWTON: I did indeed distinguish between them in nature, with gravitational phenomena involving serious mass, and inertial phenomena involving inert mass. However, the march of the stars clearly shows their mathematical equality.

NOÉ: Indeed, as long as we stay at low speeds. At very high speeds, known as relativistic speeds and which I prefer to call ultra-velocities, this equality is less true. So I defined, and demonstrated, that these two masses were mathematically distinct, albeit linked by a coefficient, the gamma factor. **The Gamma factor is the ratio of inert mass to heavy mass**[5].

$$\gamma = \frac{m_i}{m_g}$$

With γ for the Lorentz factor, m_i for inert mass and m_g for gravitational mass.

EINSTEIN: You say that this Gamma factor is the Lorentz factor. Do you mean it's the same Gamma factor we use in Relativity?

NOÉ: Yes and no! In its mathematical expression, with a square root, yes, it's the same. But in its nature, no, it's not the same at all. In Relativity theory, the Gamma factor, noted γ, is the ratio of measured time to so-called "proper" time.

EINSTEIN: Indeed, the proper time of an object is the time measured in the object's frame of reference, i.e. a frame of reference where the object is stationary.

NOÉ: A so-called proper time different from the measured time only exists if we consider that the theory of Relativity is true with its variable time depending on the reference frame. In neo-Newtonian mechanics, time flows regularly, and there is only one time, whatever the observer, as in Newtonian mechanics.

NEWTON: If we consider inertial mass to be equal to gravitational mass, then this Gamma factor is equal to unity, so it wouldn't make much difference.

NOÉ: Indeed, at normal speeds, the Gamma factor is equal to unity, so there's no difference in results. At very high velocities, such as that

---

[5] **"How to Demonstrate the Lorentz Factor: Variable Time vs. Variable Inertial Mass"**, - https://www.scirp.org/journal/paperinformation.aspx?paperid=54203

of Mercury, the fastest planet in our solar system, we can start to notice a slight difference in results between Newtonian mechanics calculations and measurement data.

EINSTEIN: This difference in results can be easily explained within the framework of relativistic theory.

THE PRESIDENT: I'm afraid we're getting a bit sidetracked here. When you sum the speed of photon emission with the speed of the reference frame, you get a photon speed greater than the speed of light. Is this neo-Newtonian mechanics?

NOÉ: Yes and no!

THE PRESIDENT: That's another Norman answer!

NOÉ: Let me explain. No, as in the theory of Relativity, there is a speed limit that cannot be exceeded. Yes, in the context of a change of reference frames, there may be cases where this speed is seen by the observer as exceeded.

GALILEO: And do you retain my Galilean principle of relativity?

NOÉ: In the same way as it is integrated into Newtonian mechanics.

EINSTEIN: Under these conditions, you retain Galileo's speed transformation.

NOÉ : Not really, because if we consider that all Galilean inertial reference frames are equivalent for describing physical phenomena, in practice, there is one preferred frame.

NEWTON: And do you retain the principle of forces at a distance?

NOÉ: Yes, even if I can't explain it any more than you can, it has to be said that they exist, whether gravitational, electromagnetic or other, in the atom. It's exactly the same situation as in Newtonian mechanics.

NEWTON: That's a shame.

THE PRESIDENT: As I've just said, the aim of these days is not to enter into debates and judgements on the two respective theories, but to remain focused on the interpretations of the experiments and observations that will be given. We have noted the fundamental difference between Newtonian and neo-Newtonian mechanics. In the former, gravitational and inert masses are mathematically equal. In the other, they are not, but are linked by a gamma factor. And, unlike Relativity, this gamma factor is not a ratio of durations, but a ratio of masses. I suggest we leave it at that for now. We'll have the opportunity to better understand the originality of this theory during the hearings of the experts we're now going to listen to.

GALILEO, *aside:* Neo-Newtonian mechanics, variable inertial masses, that's surprising too.

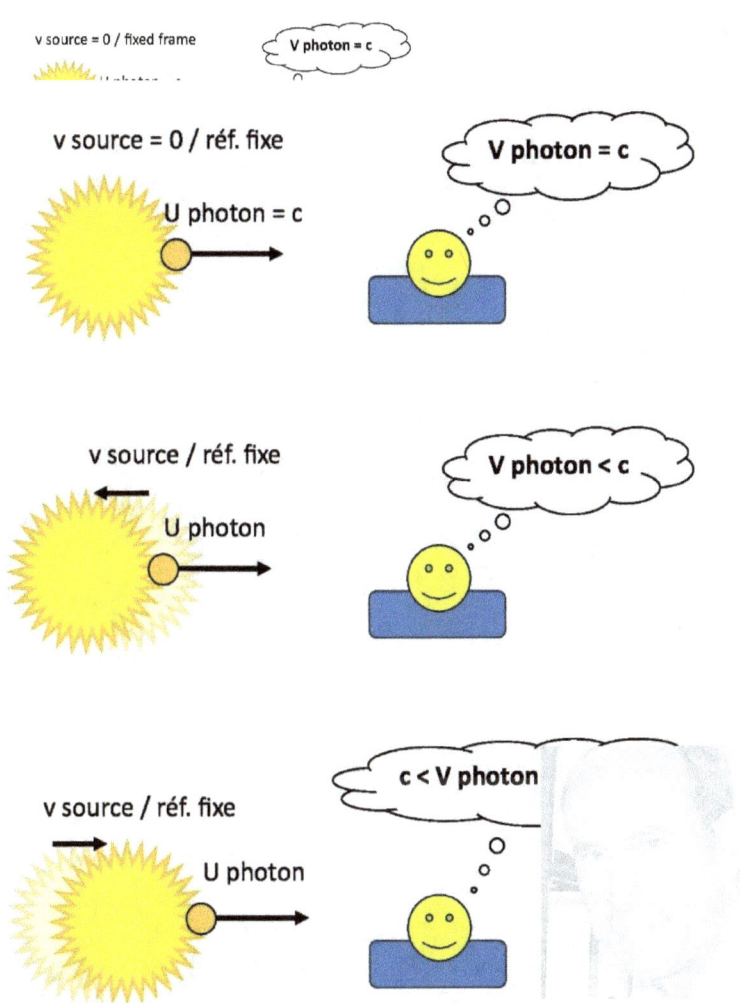

FIGURES 8: PERCEIVED LIGHT VELOCITIES

GALILEO, WAKE UP!

8. LIGHT DEFLECTION

THE SESSION PRESIDENT: Before turning to the possible evidence in favor of neo-Newtonian mechanics, we will first review the main experiments and observations in favor of Relativity. We'll set them out as factually as possible, and then discuss them according to each interpretation, the relativistic and the neo-Newtonian. To each his own! Let's start with the one that established Relativity, and which has been dubbed the first proof of Relativity. Please let the first expert in... Hello, let us introduce ourselves.

EDDIGTON: Good morning Madam President, good morning gentlemen. I'm Lord Eddington, mathematician emeritus, head of the British Royal Laboratory at Cambridge. It was I who first proved Relativity.

THE PRESIDENT: What is the purpose of your experiment?

EDDINGTON: The aim is to verify that light is being deflected by the Sun as predicted by Professor Einstein in accordance with General Relativity. This shows the curvature of space-time by matter.

THE PRESIDENT: Just to clarify the framework, we're talking about General Relativity, not Special Relativity, is that right?

EDDINGTON: As its name suggests, General Relativity is more general than Relativity Restricted to Inertial Reference Systems. Proof of one is proof of the other.

NEWTON: Concerning the deflection of light by the Sun, need I remind you that, contrary to those who claim that light is a wave, Newtonian mechanics also predicts that the light particle is attracted to and therefore deflected by the Sun?

EDDINGTON: You're absolutely right, dear Sir Newton and colleague. I'd just like to point out that the deviation predicted by General Relativity is exactly double that predicted by Newtonian mechanics. As Professor Einstein wrote, "*half of this deviation is produced by the Newtonian field of attraction of the sun, and the other half by the geometrical modification, or 'curvature', of space caused by the sun*"[6].

THE PRESIDENT: I'm not sure I understand the meaning of every word in your last sentence, especially when you talk about a "Newtonian" field in the context of Relativity. But, never mind, the point is not to get into theoretical considerations. Please describe your experience.

EDDINGTON: This involves estimating the deviation of light from the stars at the solar limb, i.e. at the periphery of the solar disk, at the time of a solar eclipse.

THE PRESIDENT: So this is a one-off observation, limited to the duration of an eclipse. It's not a laboratory experiment where you can vary the parameters, nor is it a direct measurement, since it's an estimate based, I imagine, on a mathematical model in line with relativistic theory. And what were the results?

EDDINGTON: The raw results were in fact mixed. In the observation I made at Principe in Africa, the results were in line with Professor Einstein's prediction. At Sobral in South America, under the guidance of Crommelin and Davidson, the raw results of the equivalent telescope seemed to be more in line with Newtonian mechanics alone. But this telescope was ill-suited to the humid and windy conditions at Sobral. Fortunately, the team of astronomers had also brought along a smaller spare telescope, which worked well and gave results in line with Professor Einstein's prediction.

---

[6] Relativity, Einstein, 1920, page 153, https://www.ibiblio.org/ebooks/Einstein/Einstein_Relativity.pdf

NEWTON: If I understand correctly, you have discarded the measurements of Sobral's main telescope, which gave results in line with Newtonian mechanics.

EDDINGTON: Absolutely, my dear colleague, for the reasons I've just given you. There is no manipulation here, everything is recorded in the expedition report, which has been validated by Professor Dyson, the Astronomer Royal."

THE PRESIDENT: If Davidson agreed with these conclusions based on his own measurements, one might wonder why posterity only remembered your name and not his as well, but that's a subject for historians of science. Has this observation been confirmed in other eclipse observations?

EDDINGTON: Absolutely. All subsequent observations have confirmed Professor Einstein's theory.

NOÉ: If I may say so, these results are the product of relativistic mathematical extrapolations, and the following observations were subject to significant measurement uncertainties.

THE PRESIDENT: Unless we question their probity, the fact is that they confirmed the relativistic prediction. Apart from these observations during solar eclipses, which are necessarily brief, have there been any other observations showing this deviation in line with relativistic theory?

EINSTEIN: Yes, of course! These are observations made from satellites. One example is the confirmation by the Hipparcos satellite of the relativistic factor 'PPN Gamma'. This is not to be confused with Lorentz's Gamma factor, nor with the 'Gamma C' deviation factor, which is used to quantify the difference between observation and prediction.

NOÉ: **The Hipparcos satellite didn't directly measure light deflection**[7] and this PPN Gamma relativistic factor was a bit of a snake biting its own tail. Let me explain. Let's suppose I'm a flat-earth believer. According to my Platist theory, there is a Beta factor that must be equal to unity if my Platist theory is correct. You provide me with measurements that I process according to my own theory and then, as luck would have it, I obtain a Beta value equal to one. So my flat-earth theory would be confirmed. Experimentalists proceed in the same way, grinding data according to relativistic calculations, to come up with a result that is necessarily in line with the theory of Relativity.

EINSTEIN: Under these conditions, you also reject the other 'Gamma prime' factor, which is neither the PPN Gamma factor, nor the Lorentz Gamma factor, nor the Gamma C deviation factor. You also reject all measurements made by other satellites, such as Gaia. You're closed to any explanation, young man.

NOÉ: It's rather your supporters who are closed to any possible questioning of the theory of Relativity. The **proof of this is the measurements that could be made from the Gaia satellite**[8], the communication of which would not imply any treatment, neither relativistic nor neo-Newtonian. So far, however, I have not found the operators of this satellite willing to listen.

THE PRESIDENT: Let's stop this discussion from getting out of hand. We note, Mr. Noé, that you question the relativistic mathematical treatment of satellite data. But how would you yourself interpret the results of direct measurements during an eclipse?

NOÉ: I'd like to explain it in more detail.

---

[7] "**Hipparcos did not measure directly the light bending!**", https://www.gsjournal.net/Science-Journals/Research%20Papers-Mechanics%20/%20Electrodynamics/Download/6998

[8] "**Gravitational Light Deflection: could Relativity be Invalidated by GAIA?**", http://www.mrelativity.net/Papers/51/Light%20deflection%20SERRET%20Millennium%20Aout%202018.pdf

THE PRESIDENT: Let's take a little break first. I need to collect my wits after all these different definitions of the Gamma factor.

(Further discussion in the next chapter)

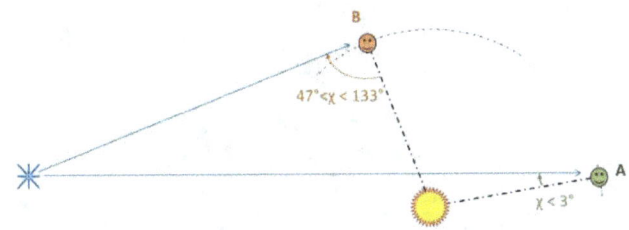

FIGURE 9: MEASUREMENT OF LIGHT DEFLECTION BY THE SUN

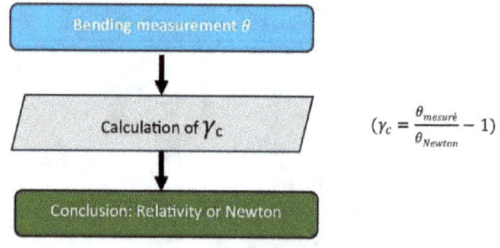
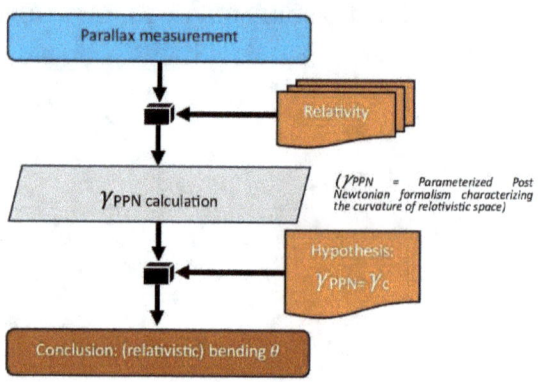

FIGURE 10: MEASUREMENT PROCESSING DIAGRAMS

9. THE SHAPIRO EFFECT

(Continuation of the chapter on light deflection)

THE PRESIDENT: Go ahead, we're listening.

NOÉ: As I was saying, direct measurements during solar eclipses have only been made with a very large measurement uncertainty. In addition to Newtonian gravitational deviation, there may be other factors disrupting the observation. Take, for example, the effect of the sun's corona, which extends over several solar radii. Its refractive power is said to be negligible, but how was this established?

EINSTEIN: On that account, we can question everything in principle, but in practice, we need more solid elements. Eddington had clearly rejected this refraction effect of the solar corona, because of its extremely low density.

NOÉ: The possible effect of refraction by gases in the solar corona is not a gratuitous hypothesis or a random question mark. Take the Shapiro effect.

EINSTEIN: If you like, but it doesn't go your way. The Shapiro effect, also known as gravitational retardation, is another phenomenon predicted by Relativity. The light's travel time is longer than expected because the photon's path is deflected and thus lengthened by the deformation of space-time.

NOÉ: The travel time of electromagnetic waves, whether light or radar, is indeed longer than expected. Whether this is due to the distortion of space-time is your interpretation. My interpretation is that the cause is the refraction generated by the gases in the solar corona.

THE PRESIDENT: We won't get you to agree.

NOÉ: But there's a simple way to tell us apart. The refraction effect of a body differs according to the wavelength of the photon passing through it.

NEWTON: Absolutely, I've made that clear. For example, by passing a ray of light through a prism, we can spread it by diffraction, generating the seven colors of the rainbow.

GALILEO: It's not anecdotal, this diffraction effect was a big problem for me when focusing my glasses.

NOÉ: An achromatic effect is when the frequency has no effect on the result. According to the theory of Relativity, since the speed of the electromagnetic wave is constant, the delay should be identical, whether it's light or a radar wave. This is an achromatic effect.

EINSTEIN: Indeed.

NOÉ: The chromatic effect is the opposite, when the frequency of the wave has an effect on the result. **According to neo-Newtonian mechanics[9] , photons of different frequencies have different energies and therefore very slightly different speeds. Thus, the delay would not be exactly the same depending on wavelength.** Unfortunately, the scientific literature available to the general public is very short on this subject.

THE PRESIDENT: What do you mean?

NOÉ: In concrete terms, one way of verifying the refraction effect of the solar corona would be to send radar waves at different frequencies towards Venus and measure, on their return, whether there is a different delay depending on the wavelength. Of course, this has already been done, but curiously enough, all that can be found in the literature accessible to the general public is that measurements of the

---

[9] **"Shapiro Time Delay derivates from Refraction",** https://www.gsjournal.net/Science-Journals/Research%20Papers-Relativity%20Theory/Download/7330

chromatic effect with waves have not been carried out with the necessary precautions. What does this tell us? Well, it's strange.

THE PRESIDENT: We take note of your interpretation of a part of the solar corona being deflected by the refraction effect, as well as your disillusioned remark, which still needs to be verified. We'll now turn to the experiments and observations specific to the second postulate of Special Relativity.

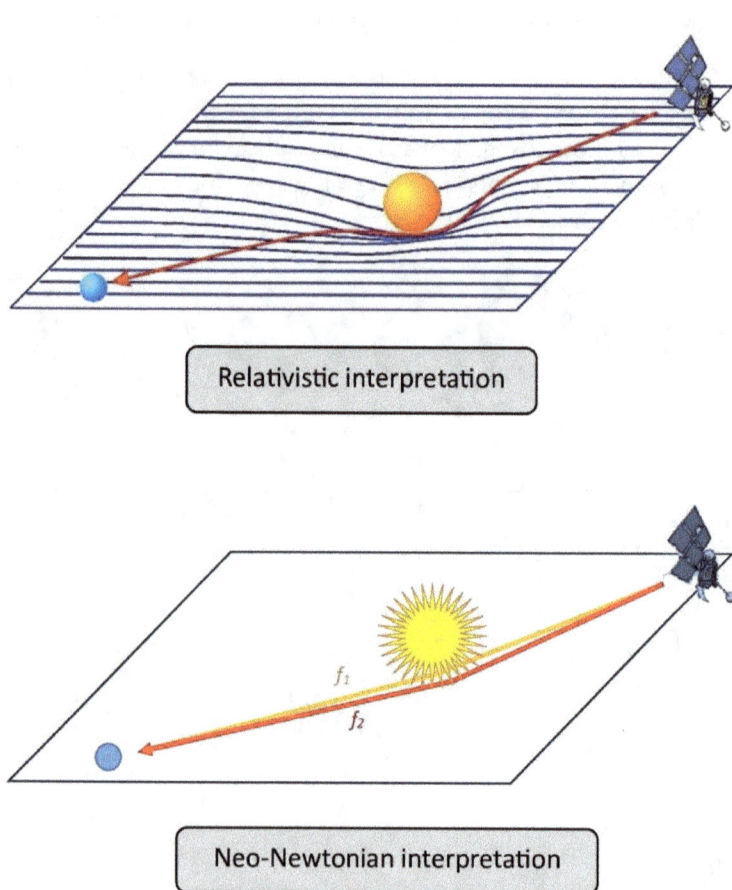

FIGURES 11: SHAPIRO EFFECT

10. THE MICHELSON EXPERIMENT

THE PRESIDENT: We continue by examining experiments and observations related to Relativity, and in particular its main postulate: the invariance of the speed of light. Let's bring in the first expert... Please introduce yourself.

MICHELSON: Good morning. My name is Albert Abraham Michelson, born in Prussia, but an American citizen and the first American to win a Nobel Prize in Physics.

THE PRESIDENT: After a Brit, an American. This shows the international nature of this commission. What have you achieved?

MICHELSON: Here's the thing. We know how to calculate the Earth's speed around the Sun. What I wanted to achieve was the measurement, not the calculation, of this speed of the Earth in space. At the time, in 1887, we thought that light moved in an ether, like a wave moves on the surface of water. By using an interferometer to send two beams of light, one in the direction of the Earth's motion, the other at right angles to it, this experiment would measure the phase difference upon arrival. This would enable us to deduce the speed of the interferometer in this luminiferous ether.

THE PRESIDENT: The link with Relativity is not obvious.

EINSTEIN: If I may say so, here it is. The result of this experiment is that no interference shift could be demonstrated, which shows the constancy of the speed of light. This is known as the "null result of Michelson's experiment", null in the obvious sense that the measurement gave a result of the order of zero.

MICHELSON: I wouldn't put it like that. The values obtained were twenty times lower than expected.

EINSTEIN: That's what I meant, with a result of the order of zero.

THE PRESIDENT: How do you interpret these very low values, Mr Michelson?

MICHELSON: The result of this experiment has been the subject of several interpretations: the total or partial entrainment of the ether, the contraction of lengths, the increase in the speed of light in the direction of propagation, the abandonment of Galileo's transformation, and, in the end, the relativistic interpretation. For my part, I refute the relativistic explanation and support the ether explanation to my dying day.

EINSTEIN: I'd like to pay tribute once again to Albert Michelson. He paved the way for the development of the theory of Relativity. Without his work, this theory would be little more than an interesting speculation. As you know, I was also able to demonstrate the photoelectric effect. This officially contributed to my being awarded the Nobel Prize, as the Nobel Committee didn't want to award it for a theory, as a theory can always be challenged in the future. But that didn't stop me from giving my prize-giving speech on the theory of Relativity alone.

THE PRESIDENT: Your article on the photoelectric effect is also theoretical: you give an explanation, but you haven't carried out any experiments. I rather think that the Nobel Committee was divided on your theory of Relativity at the time. It would be for the latter reason that they wouldn't have awarded you the Nobel Prize for it. But let's get back to Michelson's experiment and the explanation for its 'null' result.

EINSTEIN: The explanation is quite simple. The speed of light is assumed to be constant. The distances in the two arms are given to be identical, as seen by an observer linked to the interferometer, i.e. moving in space. The photons therefore arrive at the receiver at the same time. It's as simple as that, what more is there to say?

NEWTON: Yes, there is one thing, and that's the absence of a purely mathematical demonstration. You've given us a qualitative explanation, that's fine, but if we want to go further, where can we find the mathematical equations attached to this experiment?

EINSTEIN: It's true that I didn't pay too much attention to the mathematical formalism of this experiment, which is very didactic. The qualitative explanation is sufficiently eloquent on its own, so tacking on a semblance of equations won't add anything.

GALILEO: And yet, the mathematical exercise is a formidable one, revealing trickery and miscalculations.

EINSTEIN: What do you mean?

GALILEO: Nothing in particular, just a general comment I'm making to myself. However, I do have a question. According to my principle of relativity, the motion of an object in translation, i.e. in a straight line, remains unchanged if an additional parallel uniform motion is added to it. Michelson performed the experiment with an interferometer linked to the Earth and an observer linked to the Earth. **So, was this experiment repeated with the interferometer relative to the observer, or the other way round?**

THE PRESIDENT: In this case, it could be to have the interferometer linked to the Earth and the observer in space, in a rocket for example. Or the other way round, of course.

EINSTEIN: As far as I know, it hasn't been done, one way or the other. But the result would normally have been identical, in accordance with Professor Galileo's principle of relativity.

THE PRESIDENT: Let's hope so. However, only the actual realization of this other experiment, with, for example, an observer in space, would allow us to be sure. If there are no further comments on the relativistic interpretation, let's hear the neo-Newtonian interpretation.

NOÉ: Thank you, Madam Chairman, for giving me the floor. Before outlining the neo-Newtonian solution, I'd like to point out how

Michelson's experiment is historically fundamental to the imposition of the theory of Relativity. The pattern of thought is as follows. The ether hypothesis, whether partially entrained or stationary, fails to provide an acceptable or accepted explanation of Michelson's experiment. This means that the aether hypothesis is false, and that the theory of Relativity is true. This is what we call a false dilemma, or a unique choice!

THE PRESIDENT: That's your view of the matter. Please give us your interpretation of the null or near-zero results of Michelson's experiment.

NOÉ: Albert Einstein was able to show that light is made up of photons, i.e. particles. In the observer's interferometer frame of reference, it's to be expected that particles emitted at the same speed and travelling the same length should arrive at the same time.

EINSTEIN: I'm sure we agree.

NOÉ: Now let's take the case where the observer is located in Space, in a rocket as you suggested. Seen from this observer, the distances travelled by each of the two photons are different. And the speeds of the photons are also different.

GALILEO: Right! And to find the speed of the photons, you apply my principle of relativity, summing the two movements, that of the photon and that of the rocket, is that right?

NOÉ: Not in detail. Yes, I do add up the two movements, that of the photon and that of the rocket, but no, I don't do the mathematical sum of these two speeds.

GALILEO: I don't quite understand the difference you're making between movement and speed.

NOÉ: What I mean is that I'm summing up the quantities of motion, not the speeds.

NEWTON: Mathematically, it boils down to the same thing: since the masses are equal, the sum of the quantities of motion is proportional to the sum of the velocities.

NOÉ: In Newtonian mechanics, yes, quantities of motion are directly proportional to speeds. In neo-Newtonian mechanics, they are not: the Gamma factor comes into play. This is the point of my demonstration[10].

THE PRESIDENT: There's that Gamma factor again! But as I've already said, we're not here to debate theoretical calculations, but to focus on the interpretation of the experiment. We've already noted that, for this Michelson experiment, the interferometer observation needs to be performed from another frame of reference, in order to be able to conclude one way or the other. I suggest we leave it at that. Let's hear from the next expert on the same subject of the constant speed of light.

---

[10] **Which derivation for the result of the MMX in translation with respect to the observer?",** https://www.gsjournal.net/Science-Journals/Research%20Papers-Relativity%20Theory/Download/7808

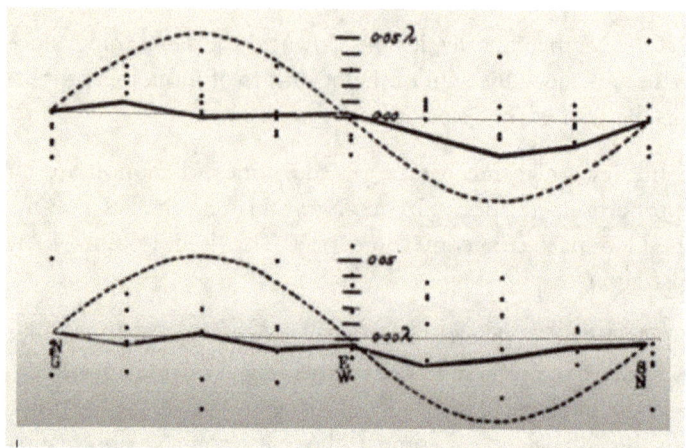

FIGURES 12: ACTUAL (SOLID LINE) AND EXPECTED (DOTTED LINE) RESULTS

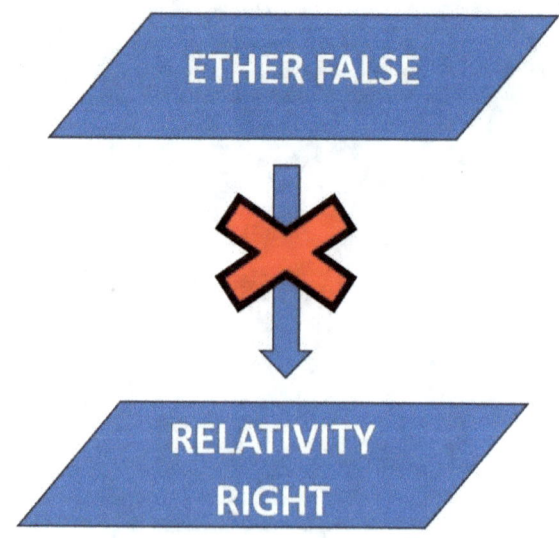

FIGURE 13: RELATIVISTIC REASONING

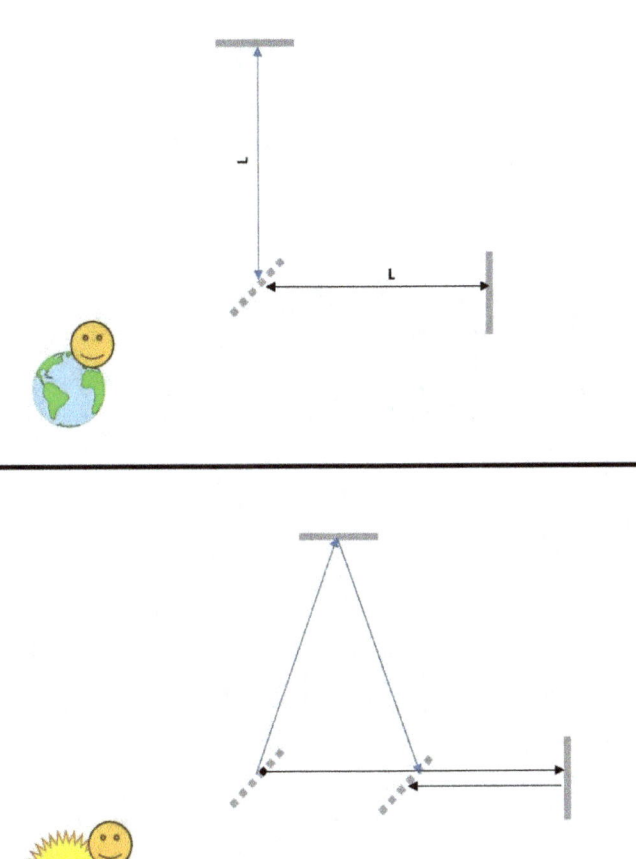

FIGURE 14: PATHS PERCEIVED FROM THE GROUND AND FROM SPACE

GALILEO, WAKE UP!

11. THE SAGNAC EXPERIENCE

THE PRESIDENT: We're continuing with our expert presentation on the main foundation of the theory of Relativity, namely the constant speed of light in any frame of reference. Can you introduce yourselves?

SAGNAC: Bonjour Madame, bonjour Messieurs. My name is Georges Sagnac, a French patriot, a fierce opponent of the ravings of Germanic physics and a proponent of the luminiferous ether.

THE PRESIDENT: Thank you. Now, a Frenchman, which, between us, confirms the international vocation of this commission. Please present your experience, avoiding partisan or nationalistic comments.

SAGNAC: It's very simple, the aim is to show that light does not travel at constant speed in any reference frame. To do this, I took a ray of light and split it in two using a semi-reflecting mirror.

THE PRESIDENT: Unless I'm mistaken, it's the same thing in Michelson's experiment.

SAGNAC: Exactly. Then I make each ray travel the same path on a disk, but with one in the opposite direction to the other. And thanks to the phenomenon of interference, the same as that used by Michelson, I see if one is behind the other. The results are as follows. When the disk is stationary, there is no phase shift. When the disk is rotating, the phase shift increases with the speed of rotation.

THE PRESIDENT: How do you analyze these results?

SAGNAC: Phase shift measures the delay of one ray in relation to the other. When the disk is stationary, the two rays have traveled the same distance, in opposite directions, but it's the same distance in length, so they arrived at the same time, and their speed was therefore identical. When the disc is rotating, there's a phase shift: they didn't arrive at the same time. From the point of view of a terrestrial observer, not linked to the rotating disc, it is clear that the distance covered is not the same. It's therefore normal for the two rays not to arrive at the same time. And if the observer is on the rotating disk, it makes no difference: the two rays still don't arrive at the same time.

THE PRESIDENT: What do you think?

SAGNAC: According to the theory of Relativity, the speed of light would be constant in any reference frame. For an observer on the rotating disk, the distance covered by the two rays is identical, so there should be no phase shift, which is not the case. For me, this experiment refutes the constant speed of light in any reference frame, and hence the theory of Relativity.

THE PRESIDENT: Here's a frontal challenge, based on the results of an experiment. How do you respond, Mr. Einstein?

EINSTEIN: Thank you, Madam President. This in no way calls Relativity into question. The second postulate clearly states that the speed of light is constant in any galilean or inertial reference frame. However, we are in the case of a rotating frame of reference, i.e. non-galilean, non-inertial.

GALILEO: I don't quite understand the difference in the behavior of light in a galilean and non-galilean frame of reference. Does this mean that the speed of light would not be constant in a non-galilean frame of reference?

EINSTEIN: It's more complex than that. You have to introduce the relativity of simultaneity. What is simultaneous in one frame of reference is not necessarily so in another. The acceleration due to the rotation of the disk is similar to gravitational acceleration. The time difference between the arrival of the light beams corresponds to the time difference between the clocks in a gravitational field. In its own frame of reference, the photon's speed is always equal to 'c', the speed of light in a vacuum, when measured in its own time. But this proper time cannot be that of the rotating disk, hence the time lag. This time dilation induces the relativity of simultaneity.

GALILEO: I modestly admit that I didn't fully understand your explanation. Right from the start, I was at a loss, because you introduced a concept I don't understand: the relativity of simultaneity. Does this mean that if two events are simultaneous in one frame of reference, they would not necessarily be simultaneous seen from another frame of reference? For example, if two balls seen from the Earth arrive at the ground at the same time, seen from the Sun, would they not arrive at the same time?

THE PRESIDENT: Professor Einstein's introduction to the relativity of simultaneity is certainly interesting, but we need an experiment to prove it. Above all, it takes us away from the current hearing on Sagnac's experiment concerning the constant speed or otherwise of light. What is the neo-Newtonian interpretation?"

NOÉ: Georges Sagnac explained it very well. **When the disk is rotating, the two beams don't travel the same distance, so it's normal that they don't arrive at the same time, whether the observer is attached to the disk or not. This is simply in line with Newtonian mechanics** [11][12]. May I add something about the relativistic explanation?

THE PRESIDENT: As long as it's brief.

NOÉ: In Sagnac's experiment, the ratio of the surface to the perimeter covered is very special, and the relativistic demonstration is based on this special ratio. In order to distinguish between the two interpretations, I propose that this experiment be repeated along different paths.

---

[11] **"A Non-Relativistic Explanation of the Sagnac Effect"**,
https://www.scirp.org/journal/paperinformation.aspx?paperid=91858
[12] https://www.youtube.com/watch?v=_FKeJ6IrmtE

THE PRESIDENT: Unfortunately, we're not in a test laboratory here, so we're evaluating the two interpretations on the basis of validated measurements, not hypothetical ones. As it stands, Sagnac's experiment shows that the speed of light does not appear to be directly constant in a rotating frame of reference, but this does not seem to directly call into question the principle of the constant speed of light in a translating frame of reference. Let's move on to the next expert.

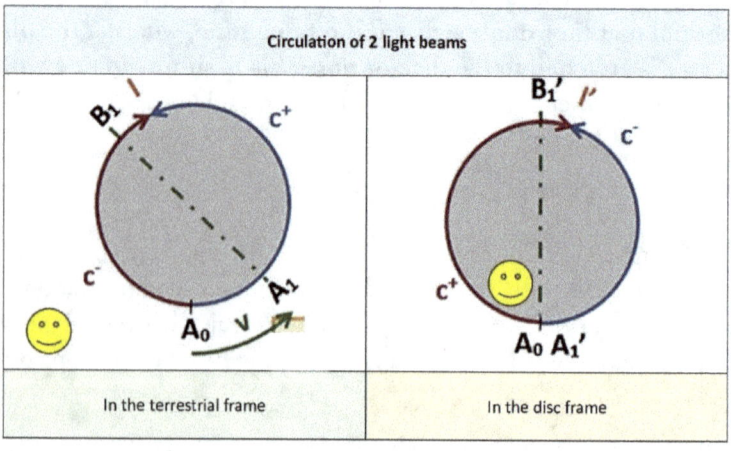

FIGURE 15: PATHS TRAVELED FROM THE GROUND AND FROM THE DISK

Configuration	Scheme	$\dfrac{4.S}{P.R}$
Current:		≈ 1
Proposed:		$\sqrt{2}$

FIGURES 16: CURRENT AND PROPOSED CONFIGURATIONS

GALILEO, WAKE UP!

12. VARIABLE DISTANCES

THE PRESIDENT: Please bring in the next expert on the constant speed of light in any inertial reference frame.

THE REPORTER: We have no further experts to present.

THE PRESIDENT: That's a shame! The first expert, Michelson, a Nobel Prize winner all the same, rejects the relativistic interpretation. The second, Sagnac, builds an experiment against the constant speed of light. Don't we have an experiment that unambiguously and uncontroversially shows the constant speed of light?

EINSTEIN: It's the very principle of a postulate: it doesn't show itself, it doesn't demonstrate itself, it admits itself as such.

THE PRESIDENT: Of course, you're right to point out this definition. But don't we have experiences that can corroborate this postulate?

THE REPORTER: It would be possible to talk about it indirectly.

THE PRESIDENT: What do you mean?

THE RAPPORTEUR: Speed is the ratio of length to time. Speed is calculated by the following mathematical formula : *speed* $c = \frac{length}{duration}$. For speed to remain constant, duration must expand and length contract.

THE PRESIDENT: I think you made a mistake when you said that duration expands and length contracts. Didn't you mean to say that, in order for speed to remain constant, duration must expand, i.e. increase, and therefore length must increase proportionally, i.e. expand and not contract?

EINSTEIN: I have to say it again. Let's get back to definitions. Time dilation is when a moving observer observes that the other person's clocks seem to slow down. Length contraction is when two observers, moving relatively to each other, measure that objects in the direction of their motion appear to shorten.

THE PRESIDENT: I'm not sure that answers my point, but the aim of these hearings is not to get into theoretical debates. *To the rapporteur.* Back to my question. What exactly did you mean by talking about it indirectly?

THE REPORTER: The proposal would be to involve witnesses on the corollary to constant speed, i.e., to avoid talking about expansion and contraction, on variable length and variable duration.

THE PRESIDENT: That's an interesting suggestion. Let's start with variable length.

THE REPORTER: There is one problem, however, and that's that I haven't found any experts on variable length. However, I've done my homework and the results seem interesting.

THE PRESIDENT: Well, go ahead, knowing that you're speaking to us on your own behalf.

THE REPORTER: Of course. Here's what I've noticed. In Relativity, there is not one distance, but at least four recognized distances to describe the same phenomenon:

- The proper distance is that between two points considered at rest in relation to each other,
- Angular distance is the apparent angle in curved space-time,
- The brightness distance is the distance that takes into account the object's brightness,
- The comobile distance is the one that takes into account the expansion of space.

Surprisingly, at high speed, these four distances do not coincide.

NOÉ: If I may say so, there's something even more surprising: the angular distance decreases as the other distances increase. That's not even surprising, it's incoherent!

EINSTEIN: The key concept in Relativity is the proper distance. I've defined it as the measure of the separation between two events in space-time. The other distances are cosmological, measuring the distance between galaxies and other cosmic phenomena.

THE PRESIDENT: It's not very clear whether the other distances are consubstantial with Relativity or additions to determine galaxy distances. But I must stop you there; cosmological distances will be dealt with when we talk about cosmology. In the interest of fairness, Monsieur Noé, can you give your definitions of lengths in neo-Newtonian mechanics?

NOÉ: Unlike the theory of Relativity, in neo-Newtonian mechanics there are not several distances for the same object depending on the observer's speed, luminosity, expansion of ...

THE PRESIDENT: Briefly, please.

NOÉ: **The distance to an object is the length that separates us from that object at the moment of its light emission or reflection.**

THE PRESIDENT: Thank you. Mr. Reporter, since we don't have any experts on variable lengths, do you have any on variable duration?

THE REPORTER: On variable duration, yes, absolutely. I'll have them call.

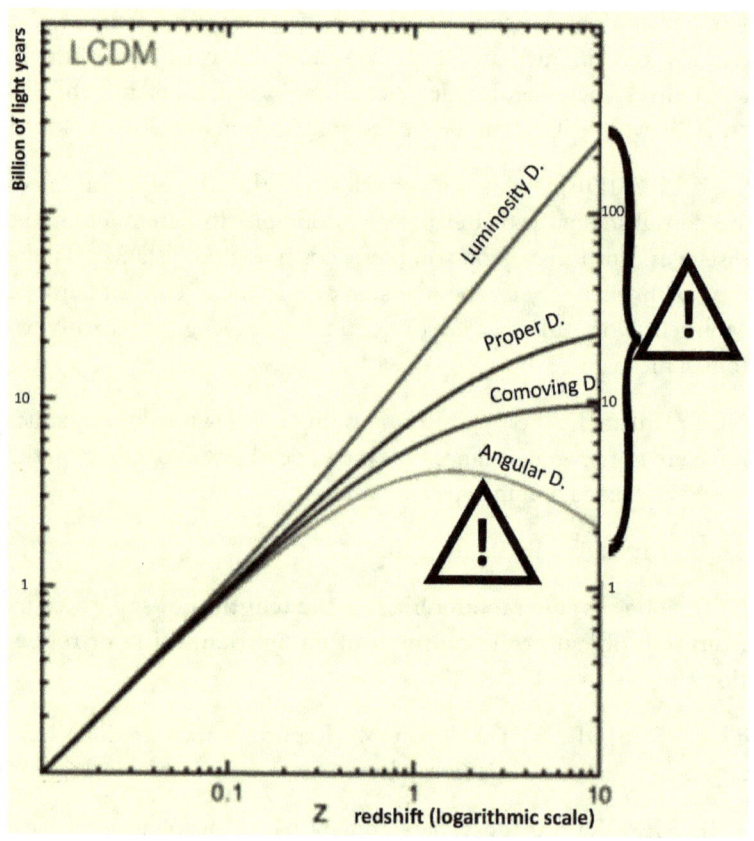

FIGURE 17: GALAXY DISTANCES IN <u>RELATIVITY</u> THEORY

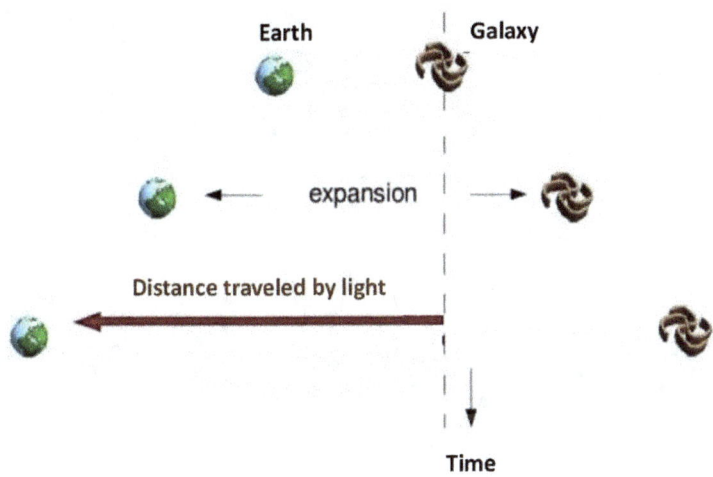

FIGURE 18: GALAXY DISTANCE IN <u>NEO-NEWTONIAN</u> MECHANICS

GALILEO, WAKE UP!

13. THOUGHT EXPERIMENTS

THE PRESIDENT: You've been called in as an expert on variable time or duration, which, along with variable length, is a corollary to the constant speed of light. Please introduce yourself.

LANGEVIN: Good morning. My name is Paul Langevin. I've discovered that the electron's inertial mass is variable as a function of its electromagnetic energy. This variable inertial mass is incompatible with Newtonian mechanics, which declares it to be constant. This is how I came to promote the theory of Relativity, which provides an acceptable framework for my discovery. I now rest in the Pantheon in Paris.

EINSTEIN, *leaning into Galileo's ear*: It's a lovely heart, by the way!

GALILEO, *answering in a low voice*: That's what makes him so likeable.

THE PRESIDENT: What's your experience?

LANGEVIN: It's a thought experiment. Imagine two twins. One travels at close to the speed of light...

THE PRESIDENT: I'm sorry to interrupt, but you say "Let's imagine". **Today we're talking about real experiences, not imaginary stories.** Theoretical suppositions will not be discussed here.

GALILEO: Madam President, allow me to say a few words. I myself have used thought experiments on several occasions. For example, the cork balls holding hands in free fall. These images have a great educational effect.

THE PRESIDENT: We have a great deal of respect for your work and your pedagogy. But fiction is fiction. It's like presenting the fictional film "Planet of the Apes" as relativistic proof; it's just an illustration, but it's not real.

EINSTEIN: Madam President, Paul Langevin didn't know it at the time, but the description of his experience was repeated by NASA's Kelly twins. One of them stayed in space for almost a year.

THE PRESIDENT: That's what I'm asking for, something concrete. What was the outcome of astronaut Kelly's stay in space?

EINSTEIN: The twin who traveled in Space had aged less, in line with relativistic calculations.

THE PRESIDENT: Good. How much has he aged less?

EINSTEIN: By 3 milliseconds.

THE PRESIDENT: Are you kidding? A lesser ageing of 3 thousandths of a second, that's ridiculous, it's surely less than its measurement uncertainty. How were those 3 milliseconds measured?

EINSTEIN: As I was saying, those 3 milliseconds weren't measured, they were calculated. But what has actually been observed is a lengthening of telomeres.

THE PRESIDENT: Could you be more specific?

EINSTEIN: Telomeres shorten with each cell division. Those of the twin who went into space had lengthened.

THE PRESIDENT: If these telomeres are longer on return than on departure, do you mean that the astronaut who spent a year in space had become younger?

EINSTEIN: No, of course not. What's more, the twin's telomeres returned to their usual size after they returned to Earth.

THE PRESIDENT: So there's been no conclusive measurement with astronauts. And this thought experiment with twins is just an imaginary story. Therefore, we won't insist, thought experiments are out of place in this evaluation commission. Please call the next expert on variable time.

GALILEO, WAKE UP!

14. VARIABLE TIME

THE PRESIDENT: Please bring in the following expert on variable time.

THE REPORTER: This is the experience of airplanes flying around the world.

EINSTEIN: Allow me to introduce it. If Langevin's twins experiment is a thought experiment, the one we're about to see is quite real.

THE PRESIDENT: Thank you, Mr. Einstein, but it wasn't necessary. It's a famous experiment. Two jets took off, each with a clock on board. One headed east, the other west, around the Earth, where they eventually returned to their starting point. The travel times could then be compared, and found to be within 10% of relativistic predictions.

NOÉ: I'm sorry, Madam President, but that's not how it happened! The reality is different. First of all, there were not one but four clocks on board. Above all, they were taken on different commercial flights, with some fifteen stopovers and as many accelerations and decelerations, first to the east and then, in a staggered fashion, to the west. The various raw results had to be reprocessed before they could be compared with a set of clocks left on land.

THE PRESIDENT: Um, we'll see about that. Please introduce yourself, Dr. Hafele.

HAFELE: Joseph Carl Hafele, American, PhD in nuclear physics, teacher at NASA.

EINSTEIN: You've come a long way, my fellow countryman!

NOÉ: After your famous experiment, however, you weren't celebrated by the relativist community and went off to work in construction, is that right?

THE PRESIDENT: This ad personam insinuation is unbearable, you're trying to devalue this expert. We're interested in experiences, not people. Please continue, Dr. Hafele. First of all, what is the purpose of this experiment?

HAFELE: The aim of this experiment was to measure the speed of moving clocks compared with stationary clocks. Contrary to what is sometimes imagined, the slowing down of time as a function of speed is slight compared with the acceleration of time as a function of the gravitational effect and the variable influence of the Sagnac effect.

THE PRESIDENT: So you're saying that the slowing down of time as a function of speed is a minority effect. I don't quite understand the point of your statement, which is supposed to prove that time slows down with speed. Please start by describing your experiment.

HAFELE: This experiment is not mine alone; I did it jointly with my colleague Keating, a timing specialist at the US Naval Observatory. But it is in line with the description just given by Monsieur Noé, concerning the use of commercial flights and the presence of four clocks on board.

THE PRESIDENT: And what was the result of the experiment?

HAFELE: In accordance with Mr. Einstein's theory.

EINSTEIN: Thank you. What Dr. Hafele didn't mention was that this experiment was carried out with atomic clocks of unprecedented precision.

THE PRESIDENT: That doesn't tell us anything about their reliability. Has this experiment been repeated by independent teams?

HAFELE: It's been redone. Twenty-five years later, we ourselves flew back and forth between the United States and England, with the same result.

NOÉ: That's not what you call going around the world again.

THE PRESIDENT: On this point, I agree with you, it's not what you'd call redoing the experiment, nor having independent teams. Mr. Noé, you have the floor. What is your interpretation of this result, which is of some interest, given that it still conforms to the relativistic prediction?

NOÉ: First of all, let's return to the subject of the experiment. According to the theory of Relativity, the influence of speed in this case would be a minority, less than 15%. What's the point of citing this experiment as proof of variable time?

THE PRESIDENT: Any other comments?

NOÉ: Secondly, the official results communicated are a specific synthesis of measurements whose reprocessing rules are not given in the report. And, for some reason, the raw results of the measurements have long been kept secret, even classified as defence confidential to be precise. It makes you wonder why. And **when you examine these raw measurements, they bear little relation to the official results.** [13][14]

THE PRESIDENT: So, the raw results of this famous but unique experiment are not as conclusive as I'd imagined. We won't examine the cause of the difference between the raw results and the official result, which is beyond the scope of this commission, and we'll leave it at that for this experiment.

---

[13] https://www.youtube.com/watch?v=myG9j35jcjs
[14] https://www.youtube.com/watch?v=uTeVCZIh8gM

GALILEO: As long as it hasn't been carried out in a laboratory where the various parameters can be controlled and varied, can we really call it an experiment?

THE PRESIDENT: I don't know. Let's listen to the next expert.

Clock #	East	West
# 120	-490	-300
# 361	120	510
# 408	50	490
# 447	-565	-550
Average	**-221** (± 344)	**38** (± 588)
Official R.	-59	273
Theoretical R.	-40	275

FIGURE 19: WORLD TOUR RESULTS TABLE

15. THE MUON EXPERIMENT

THE REPORTER: We have one last experiment on variable time, the muon experiment.

THE PRESIDENT: We're listening, sir.

SMITH: Good morning. My name is James Hammond Smith, a Harvard graduate and professor at the University of Illinois. I'm the author of an introductory book on Relativity.

EINSTEIN: An excellent book, by the way.

THE PRESIDENT: Thank you, Mr. Einstein. Professor Smith, could you explain the purpose of your experiment?

SMITH: Of course. It's an experiment I conducted jointly with Professor Frisch of MIT. The aim was to measure the lifetime of unstable muon particles in the atmosphere.

THE PRESIDENT: What exactly does the experiment involve?

SMITH: When muons decay in the scintillator, or die as it were, they cause the emission of photons of light, or scintillations, which can be detected. By measuring the time between the arrival of a muon in the scintillator and the moment when the scintillations are detected at two different altitudes, it is possible to determine the average lifetime of muons before they disintegrate. We can also determine the number of muons lost to decay between these two different altitudes.

THE PRESIDENT: What was the result?

SMITH: Let's remember that half-life is the time it takes for half of a muon population to decay. The result of the measurement is that muons have a very short half-life, of the order of two microseconds.

EINSTEIN: And the consequence of this result is that this very short time would not have allowed them, according to Newtonian mechanics, to cover the distance between the two altitude levels. Only the time dilation of Relativity can explain why muons travel such a distance in such a short time.

THE PRESIDENT: What do you mean?

EINSTEIN: These muons travel at practically the speed of light. In accordance with Relativity, at these relativistic speeds, we can consider that either the muons' time has dilated, or the distances they travel have contracted. In other words, only Relativity can explain why muons travel such a great distance in such a short time.

THE PRESIDENT: It seems the best has been saved for last. This is clearly an experiment in favor of variable time and even variable distances. If you have another explanation, we'd love to hear from you, Mr. Noé.

NOÉ: Thank you, Madam President. Indeed, as Messrs Smith and Einstein explained, this experiment was designed to prove the theory of Relativity. However, there is **another interpretation, that of neo-Newtonian mechanics, which is that the lifetime of muons at rest is not the same as that of muons at ultra-high velocities**[15] **. To be more precise, according to neo-Newtonian mechanics, the half-life of muons depends on their energy.** Proof of this is provided by CERN's measurement of the half-life of muons at ultra-high velocities. The measured half-life is of the order of several tens of microseconds, thirty times longer than the half-life of muons at rest. This explains why they can travel distances thirty times greater.

EINSTEIN: What you're really saying is that radioactive half-life depends on the speed of unstable particles. That's a completely far-fetched idea.

---

[15] **"Muon Lifetime would depend of its Energy"**, http://www.mrelativity.net/Papers/51/Muons%20Serret%20Millennium.pdf

NOÉ: That sounds less far-fetched than saying that time itself would depend on the observer's speed, and that there would be as many time flows as there are observers.

THE PRESIDENT: Come, come, gentlemen. It's been a long day, we've held a dozen hearings, and now we're going to close this first day on the foundations of Relativity. Admittedly, these experiments and observations in favor of Relativity, whether on the deviation of light, the constant speed of light or the variation of time, are not all proving as convincing as I'd imagined. That's all for today's audition. Have a good rest. Tomorrow, we'll be looking at experiments and observations more related to Neo-Newtonian Mechanics.

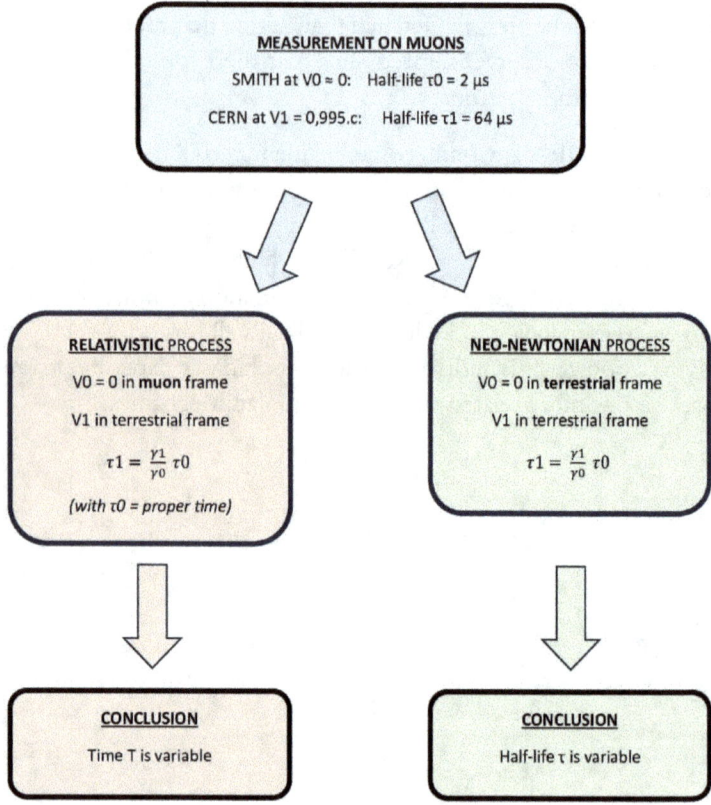

FIGURE 20: RELATIVISTIC AND NEO-NEWTONIAN MEASUREMENT PROCESSING

DAY TWO (NEO-NEWTONIAN)

NEO-NEWTONIAN EXPERIMENTS AND THEORY

GALILEO, WAKE UP!

16. THE MICROSCOPE EXPERIMENT

THE PRESIDENT: Good morning. After a first day of hearings focused more on Relativity, we start this second day with experiments and observations more directly related to hypothetical neo-Newtonian mechanics. Neo-Newtonian mechanics is based on the hypothesis that inertial mass is quantitatively different from gravitational mass. Send in the first expert.

THE REPORTER: Unfortunately, we don't have a representative to explain the following experiment, which is a joint effort, but I can give you a brief presentation.

THE PRESIDENT: It's a bad omen, but go ahead!

THE REPORTER: This is the MICROSCOPE experiment. Its aim is to verify the principle of mass equivalence by measuring the relative acceleration between two different masses placed in orbit around the Earth.

THE PRESIDENT: The principle of mass equivalence is to Relativity what the equality of masses, inert and serious, is to Newtonian mechanics. Can you describe this experiment?

THE RAPPORTEUR: This measures the energy required to hold two different masses side by side inside a satellite. These two masses, like the satellite, are in free fall.

NEWTON: In a way, it's like the Moon. It's in free fall, but because of its tangential speed, we don't realize it.

THE PRESIDENT: "*You'd have to be Newton to see that the moon is falling, when everyone else can see that it's not,*" wrote Paul Valéry. That was an aside.

KEPLER: Since orbits are in fact elliptical rather than circular, this means that half the time the satellite is descending relative to the Earth, and half the time it's ascending relative to the Earth.

GALILEO: A falling object that rises is counter-intuitive, to say the least! It's quite a change from the free fall of two balls, whether from a bell tower or even the Leaning Tower of Pisa.

THE PRESIDENT: You all seem to be questioning the very principle of this MICROSCOPE experiment. What were the results of the measurements?

THE REPORTER: The results are given in the form of the Eötvös coefficient. This involves comparing the ratios of gravitational mass to inertial mass for each of the two masses.

THE PRESIDENT: This coefficient comes from Baron von Eötvös, who compared serious and inert masses on a torsion balance and concluded that they were equal. What about the MICROSCOPE experiment?

EINSTEIN: To answer Madame President's question, the gap was again zero, which confirms Relativity.

NOÉ: More to the point, the Eötvos coefficient doesn't disprove it. Even the slightest deviation would have disproved the theory of Relativity, which makes no distinction between inert and gravitational mass. For the same reason, it would have undermined Newtonian mechanics, which was not the case. On the other hand, since **the Eötvös coefficient consists solely of the difference between the deviations, this experiment also works in favor of neo-Newtonian mechanics. Indeed, even if the ratio between inertial mass and gravitational mass is different from unity for each of the two bodies, subtracting these ratios cancels out the difference.**

THE PRESIDENT: In short, this experiment does not appear to discriminate between the two theories to be examined by this commission.

NOÉ: In fact, the two theories give fairly similar results, but their basic principles are very different. For one, the duration is variable, for the other, it's the inertial mass which is variable. This leads to very different interpretations.

THE PRESIDENT: Well noted. Let's hope that the next experiment will be more discriminating between the two theories.

$$\delta = 2\,\frac{\dfrac{m_g(A)}{m_i(A)} - \dfrac{m_g(B)}{m_i(B)}}{\dfrac{m_g(A)}{m_i(A)} + \dfrac{m_g(B)}{m_i(B)}}$$

FIGURE 21: EÖTVÖS COEFFICIENT

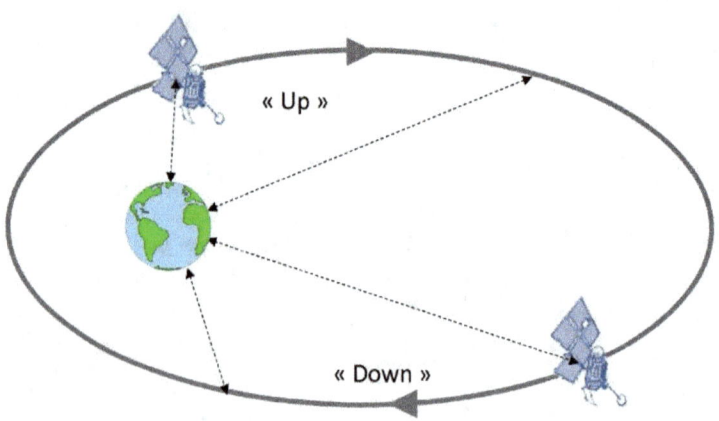

FIGURE 22: FREE FALL OF A SATELLITE

17. BERTOZZI'S EXPERIENCE

THE PRESIDENT: Please bring in the next expert.

BERTOZZI: Hello. My name is William Bertozzi, professor at MIT in Boston, USA.

THE PRESIDENT: What is your experience?

BERTOZZI: The experiment consists in accelerating electrons at very high electrical voltages.

THE PRESIDENT: That's succinct and clear. And what was the result?

BERTOZZI: The result is that at very high voltages, this speed saturates at the speed of light "c" and never exceeds it, contrary to Newtonian mechanics.

EINSTEIN: This is an excellent experiment in favor of Relativity.

NOÉ: Neo-Newtonian mechanics also recognizes this limit.

THE PRESIDENT: How do you explain the result, Professor Bertozzi?

BERTOZZI: My aim was not to confirm Relativity, but to gain a better understanding of atomic phenomena. What I found was that the ratio of the kinetic energy supplied to the velocity obtained squared was proportional to mass, according to Lorentz's Gamma factor.

THE PRESIDENT: There's that famous Gamma factor again.

NOÉ: **The result of this experiment confirms the Gamma ratio between inert mass and gravitational mass. The result clearly**

supports neo-Newtonian mechanics[16]. Indeed, to a lesser extent, it is in favor of the theory of Relativity, since in the latter, the Gamma ratio is a ratio of durations, not masses.

EINSTEIN: Relativity also recognizes the application of the Gamma factor to masses.

THE PRESIDENT: This is becoming a theoretical debate. *Rustling in the room.* In any case, it evokes the variable inertial mass of the electron established by Professor Langevin.

NOÉ: Indeed, but Professor Langevin only turned to Relativity by default, to have a theoretical framework for his discovery.

THE PRESIDENT: Has this experience been repeated by other teams?

BERTOZZI: Probably, but I don't know.

THE PRESIDENT: If this experiment was indeed carried out in the laboratory, it needs to be repeated by independent teams if it is to be valid. If the result is confirmed, this speed limit does indeed invalidate Newtonian mechanics. But this would not allow us to discriminate between neo-Newtonian mechanics and the theory of Relativity, which also seems to recognize the variability of mass. The hearing was brief, but to the point.

---

[16] **"How to Demonstrate the Lorentz Factor: Variable Time vs. Variable Inertial Mass"**, https://www.scirp.org/journal/paperinformation.aspx?paperid=54203

$$\begin{cases} \text{In \textbf{Relativity} theory,} & \gamma = \dfrac{\Delta t \text{ (measured time)}}{\Delta \tau \text{ (proper time)}} \\ \text{In \textbf{neo} – \textbf{Newtonian} mechanics,} & \gamma = \dfrac{m_i \text{ (inertial mass)}}{m_g \text{ (gravitational mass)}} \end{cases}$$

FIGURE 23: RELATIVISTIC AND NEO-NEWTONIAN DEFINITIONS OF THE GAMMA FACTOR

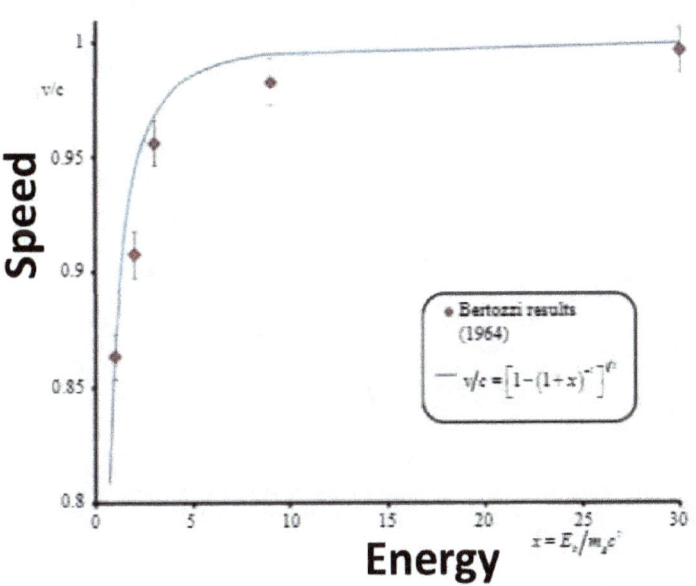

FIGURE 24: BERTOZZI'S RESULTS

GALILEO, WAKE UP!

18. THE FIZEAU EXPERIMENT

THE PRESIDENT: This experiment is generally given as a proof of Relativity, especially for the relativistic addition of speeds. But we'll also hear the neo-Newtonian interpretation, and how it differs from the relativistic one. We'll leave you to introduce yourself.

FIZEAU: Good morning Madam President, good morning Gentlemen. My name is Armand Hippolyte Fizeau, a physicist specializing in light phenomena. I measured the speed of light, demonstrated the eponymous 'Doppler-Fizeau' effect, and verified the addition of speeds according to Fresnel's coefficient.

THE PRESIDENT: What effect will you be developing on the theory of Relativity?

FIZEAU: To tell you the truth, I don't know.

EINSTEIN: Allow me to intervene. We're talking here about the moving water experiment, which is the experiment that validates the relativistic addition rule for velocities. This experiment predates the development of the theory of Relativity, hence this answer. However, this experiment decides in favor of the equation deduced from Relativity, and even in a very exact way. I'll leave it to you to present it.

FIZEAU: As you wish. The purpose of the moving water experiment is to verify the accuracy of Fresnel's coefficient, which is used to explain the addition of light-wave velocities in the ether.

THE PRESIDENT: I still don't see the connection with the theory of Relativity, but please describe your experience.

FIZEAU: I don't see the connection either. The experiment consists in measuring, with the help of interference, the difference in speed between a light beam moving in the direction of the water current and another light beam moving against the direction of the water current.

THE PRESIDENT: And what was the result?

FIZEAU: I obtained exactly the value deduced from the Fresnel coefficient.

THE PRESIDENT: Has this laboratory experiment been reproduced by independent teams?

EINSTEIN: Absolutely. For example, Michelson's study in the 19th century found this result to within +/-5%, and Toulouse's study in the 21st century also found this result to within +/-8%.

THE PRESIDENT: It's rather curious. The more we repeat this experiment with modern means, the more inaccurate it seems to become.

EINSTEIN: These are normal levels of uncertainty, even in the laboratory. The important thing is that the result invalidates the Newtonian addition of velocities.

GALILEO: That is, it invalidates Galileo's law of velocity transformation, if I understand correctly.

THE PRESIDENT: I'm well aware that this is not an observation, but a laboratory experiment, with all the parameters under control, and reproduced over the centuries by independent teams. But what does this have to do with the theory of Relativity?

EINSTEIN: It turns out that the Fresnel coefficient measured by Monsieur Fizeau is a special case of the Gamma coefficient of Relativity. The result of this experiment thus confirms Relativity.

THE PRESIDENT: Ah, there's that famous Lorentz Gamma factor again. And what is the neo-Newtonian interpretation of this result?

NOÉ: Before I give you my interpretation, let me give you some background on the development of the theory of Relativity.

THE PRESIDENT: No, the point is not to discuss the theory itself, or even its history, but to focus on the facts, the experiences.

Murmurs in the room.

NOÉ: Good Madam President. As you pointed out, the results of Fizeau's experiment clearly invalidate the Newtonian addition of velocities. Let's just point out that velocity is not, strictly speaking, a physical concept directly accessible to measurement.

THE PRESIDENT: What do you mean?

NOÉ: Let me explain. Speed is the ratio of a distance covered to a given time. It is therefore the result of a calculation. **In neo-Newtonian mechanics, we are currently hesitating between two methods of calculating the addition of velocities, one based on momentum[17] and, alternatively, the other based on kinetic energy[18].** The first calculation gives a result very close to that predicted by the Fresnel coefficient, while the second is identical."

$$\begin{cases} Momentum : \gamma_u . u = \gamma_v . v + \gamma_w . W \\ Kinetic\ energy : u = v + W/\gamma^2 \end{cases}$$

---

[17] **"Velocity Addition Demonstrated from the Conservation of Linear Momenta, an Alternative Expression"**, https://www.scirp.org/journal/paperinformation.aspx?paperid=56126

[18] **"The new speed composition u=v+w/G² in agreement with Fizeau's experiment",** https://hal.science/hal-02508527/file/HAL%20SERRET%2014%20mars%202020.pdf

EINSTEIN: Except that the result of Fizeau's experiment corresponds exactly to Fresnel's prediction and therefore to the relativistic prediction.

NOÉ: To the nearest measurement uncertainty. The origin of this uncertainty lies mainly in the turbulence of the moving water. This is **why we propose to repeat this experiment with a rotating glass disk, in order to avoid the turbulence of moving water[19] . This would even make it possible to discriminate between the two theories**.

THE PRESIDENT: And between your two neo-Newtonian formulas for adding speeds, if I've understood correctly.

NOÉ: Yes, it's the experiment alone that will decide. To come back to the correspondence between Fizeau's experiment and the theory of Relativity, it's normal, because the theory of Relativity derives from Lorentz's equations with their Gamma factor, and these equations derive from Fizeau's experiment and therefore ...

THE PRESIDENT: Please, no historical assumptions. The purpose of these days is not to discuss theoretical concepts endlessly. What we do know is that, with the experiment you propose involving a rotating glass disk rather than turbulent moving water, there may at last be a way of discriminating between the two theories. Thank you very much.

---

[19] **"An improvement of the accuracy of Fizeau's experiment"**, https://www.gsjournal.net/Science-Journals/Research%20Papers-Relativity%20Theory/Download/7247

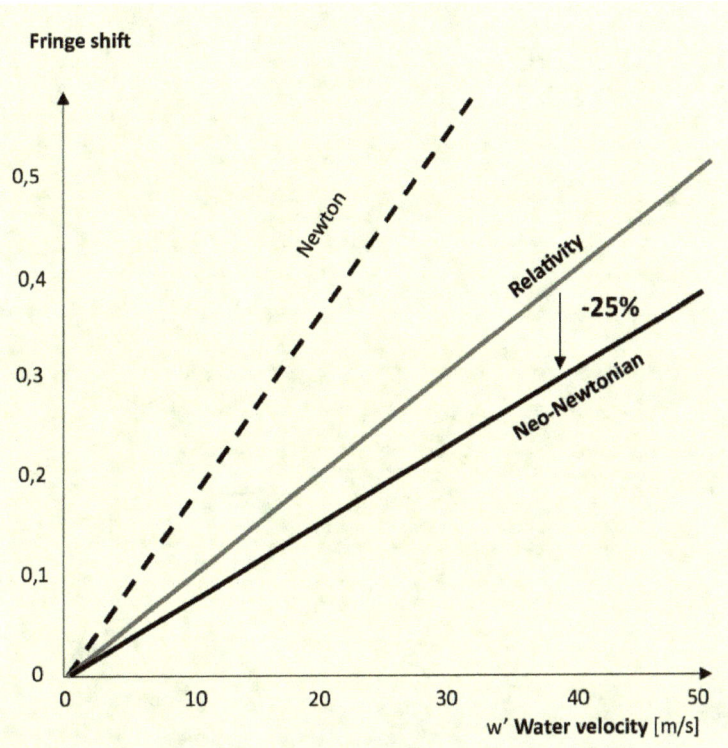

FIGURE 25: FRINGE DISPLACEMENT AS A FUNCTION OF WATER VELOCITY

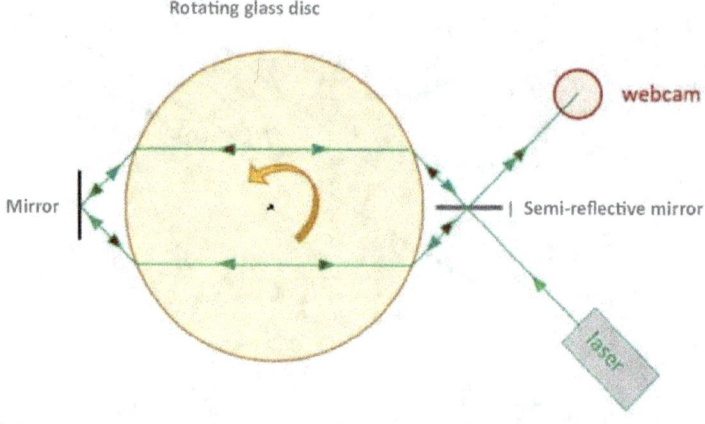

FIGURE 26: PROPOSED EXPERIMENT WITH A GLASS DISK

19. THE GPS

THE PRESIDENT: Let's bring in the next expert.

THE REPORTER: There's no particular expert, because it's a collective work. So I propose to give you a presentation.

THE PRESIDENT: Decidedly! But we're listening. What's it all about?

THE REPORTER: It's all about GPS. As an oft-used expression goes, "Without Relativity, there's no GPS!"

THE PRESIDENT: Let's avoid unauthorized quotes, please. Please remind us what GPS is and how it works.

THE REPORTER: GPS is a satellite navigation system used to determine the precise position of a GPS receiver, such as a cell phone or vehicle navigation system. It operates via a network of satellites orbiting the Earth. Each GPS satellite transmits signals containing its orbital position and the time of transmission. The GPS receiver receives these signals and processes them by trilateration to determine its geographical position. The more signals the receiver receives from different satellites, the more precise its location.

THE PRESIDENT: What is trilateration?

THE REPORTER: Trilateration is similar to triangulation. Triangulation, once used by ships' captains to take stock of their position, involves taking bearings from at least three lighthouses in order to situate oneself in the 2D plane. The more lighthouses are surveyed, the smaller the triangle of uncertainty. Trilateration, used by GPS, consists of measuring the distances of at least four satellites in order to locate in 3D space. The more satellites are tracked, the smaller the uncertainty. Triangulation is a point with directions; trilateration is a point with distances.

THE PRESIDENT: How accurate are these measurements?

THE REPORTER: For civilian GPS, it's on the order of a metre or, at worst, ten metres.

THE PRESIDENT: How does GPS prove the theory of Relativity?

EINSTEIN: Let me take over the theoretical explanation. According to Relativity, there are two opposite effects. The first, according to Special Relativity, tends to slow down time. The second, according to General Relativity, tends to speed it up. This is what we saw with the airplane experiment.

THE PRESIDENT: So?

EINSTEIN: If the mathematical models didn't compensate for this, we'd end up with an error of 11,000 m per day.

THE PRESIDENT: That's impressive! It reduces uncertainty by more than 1,000 times, from 11,000 m to around ten meters of localization uncertainty.

NOÉ: What's impressive is that it's an urban legend that's been widely circulated, but it's not true.

THE PRESIDENT: What do you mean?

NOÉ: There are **many sources of disruption to these GPS clocks, such as vibrations, altitude differences and so on. That's why, every day, the satellites are actually resynchronized by the ground control segments, without the need for relativistic correction**.

EINSTEIN: The fact that clocks are resynchronized every day doesn't prevent the use of relativistic formulas.

Murmurs in the room.

NOÉ: As they are resynchronized, it makes no difference whether relativistic formulas are used or not.

THE PRESIDENT: Where does this "urban legend", as you call it, come from?

NOÉ: When one of the test satellites, NTS-2, was originally developed, relativistic equations were used, but this was not enough to resolve the large measurement uncertainties.

THE PRESIDENT: There's no evidence that these relativistic equations are no longer in use.

NOÉ: The details of the computer program used to operate the GPS are a military and industrial secret. It's not a transparent experiment that can be reproduced by independent teams. The same will apply to the European Galileo system. As a result, GPS cannot honestly be classified as a practical application of Relativity.

THE PRESIDENT: Until proven otherwise, GPS incorporates relativistic equations into its calculations. This is practical, observable proof of the effects of both Special and General Relativity.

Hubbub in the room.

A VOICE: It's unverifiable! How can you accept so-called unverifiable proof?

ANOTHER VOICE: All we talk about here is evidence, but never theory! Why not discuss theoretical inconsistencies?

ANOTHER VOICE: Quantum mechanics, with its constant flow of time, is not compatible with the variable time of Relativity. Our real world cannot be subject to two different laws of physics!

THE PRESIDENT: Please, the floor is not yours.

A VOICE: According to Relativity, there are no forces at a distance, only deformations of space-time. How then do we explain the effect of electromagnetic forces at a distance?

GALILEO, WAKE UP!

ANOTHER VOICE: An Earth that moves straight through space-time is as absurd as a flat Earth!

ANOTHER VOICE: And this criterion of space-time with a duration equivalent to a length, it's idiotic!

THE PRESIDENT: That's enough, this is not a marketplace, you're forcing me to suspend the meeting! Please clear the room, while we wait for the serenity required for the expert hearings to return.

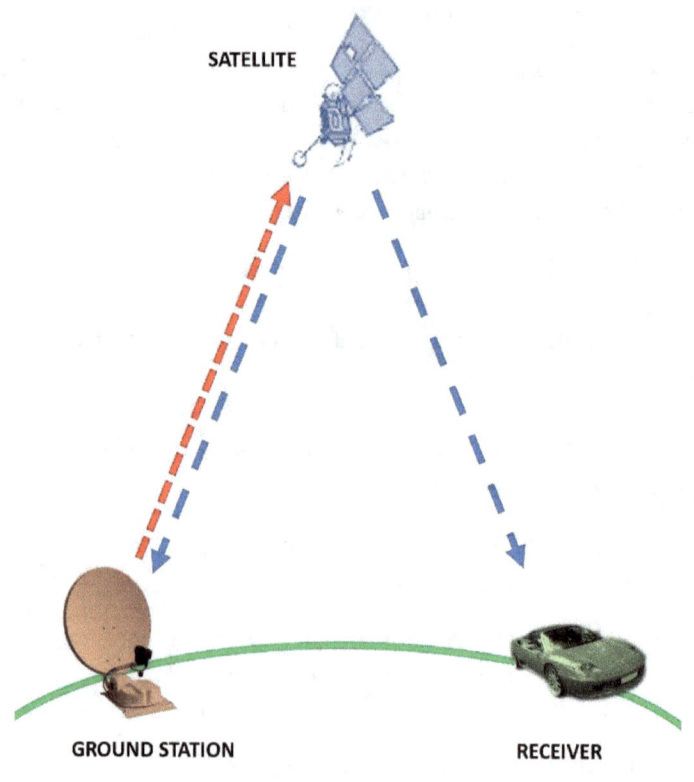

FIGURE 27: GPS SATELLITE RESYNCHRONIZATION

20. INTERMEDIATE

DISCUSSION ON THE FIRST POSTULATE

GALILEO: Finally, if today is going to be as full of hearings as yesterday, this interruption won't hurt.

KEPLER: What do you mean?

GALILEO: This break in the auditions allows us to take stock. In each of the two theories, Relativity and Neo-Newtonian Mechanics, they recognize my principles.

KEPLER: While I don't dispute them, no expert has come forward with any experience to back them up.

GALILEO: That's the very definition of a principle: it can be accepted, it can't be demonstrated.

KEPLER: But it can be seen.

GALILEO: That's what I did. I had observed that when an object slides on a smooth, horizontal surface, it continues to move without needing a continuous force to keep it in motion, hence my Galilean principle of inertia[20].

NEWTON: I took it up again in my first law[21], and the associated frame of reference is the inertial or Galilean frame of reference.

---

[20] Galilean principle of inertia: "*Every body perseveres, in a state of rest or uniform motion, in a straight line unless acted upon by external forces*".

[21] Newton's law of inertia: "*An object subjected to zero or counterbalancing forces is either stationary or in uniform rectilinear motion*".

GALILEO: I've also given my name to a type of repository, which is quite flattering.

KEPLER: What is a Galilean frame of reference?

NEWTON: A galilean frame of reference translates rectilinearly and uniformly with respect to another galilean frame of reference.

KEPLER: Defining a so-called Galilean frame of reference from another so-called Galilean frame of reference doesn't seem to me to advance the case, but rather makes it go round in circles, doesn't it?

NEWTON: You're right. I meant that a galilean frame of reference is one in which my first law is true.

KEPLER: So, for you, a Galilean frame of reference is one in which the Galilean principle of inertia is true. But in the theory of Relativity, it's the one where the Galilean principle of relativity is true. Not quite the same thing, is it?

GALILEO: Indeed. On reflection, what bothers me most is that the first postulate of the theory of Relativity[22] is a little different from my principle of relativity[23]. The first postulate speaks of identical laws of physics, whereas I only speak of the motion of translating objects.

KEPLER: It's a pity that neither the formulations, nor any associated experience, have been examined by this commission. It's a good idea for us to discuss it amongst ourselves. And what do you think of his other postulate?

DISCUSSION ON THE SECOND POSTULATE

---

[22] First postulate of the theory of Relativity: "*The laws of physics are the same for all inertial observers*".

[23] Galilean principle of relativity: "*The motion of a translating object remains unchanged if an additional uniform motion is added to it*".

GALILEO: If I've got it right, the second postulate of Relativity states that the speed of the photon is invariant in any Galilean, not to say inertial or relative, reference frame.

KEPLER: Does this mean that this speed is variable in other reference frames?

GALILEO: It's not very clear. That's why I don't know how to take Sagnac's experiment with the rotating disk, where the two photons didn't arrive at the same time. A priori, travelling the same distance without arriving at the same time hardly proves constant speed in a rotating frame of reference.

KEPLER: Relativists do explain it, though, which is surprising. After all, what difference does it make whether a frame of reference rotates or not?

GALILEO: I don't know. As for the uniform translation reference frame, there's Michelson's experiment. But, in discussions, this experiment has been denounced as incomplete, because it was not measured from another Galilean reference frame.

NEWTON: What surprised me was the law of addition of speeds. One plus one would no longer make two.

GALILEO: But Fizeau's experiment with moving water is quite clear. The law of addition of relativistic speeds is not linear. Einstein explained this to me using the example of automobiles. Because nothing can go faster than light, it's a speed limit.

KEPLER: The existence of a limiting velocity does not necessarily mean that the displacement of a light photon has the same velocity from a moving reference frame as from a stationary one. We're talking about two different concepts here.

GALILEO: Experimental evidence of the invariance of the speed of light seen from any reference frame, or at least a Galilean one, is not so clear-cut after all.

DISCUSSION OF TIME AND VARIABLE DISTANCES

KEPLER: What I really appreciated were the corollaries of this second principle, even if they are still surprising. For example, if speed is constant, then distances and durations are variable.

GALILEO: On variable distances, there have been no proposed experiments, except perhaps with muons.

KEPLER: On the other hand, I didn't understand the four types of distance. I'm looking forward to what will be said at the cosmology hearings.

GALILEO: Yes, indeed. As for variable time, there's the experience of commercial aircraft flying around the world. But it's only been done once.

KEPLER: And the raw results have, it seems, been heavily reworked. Under these conditions, it's difficult to accept the conclusions. As for Langevin's twins' thought experiment, it's not an experiment, it's just a nice story.

GALILEO: Don't be so harsh, I too have used images and thought experiments. You too have thought the world through a model.

KEPLER: Absolutely, but a model remains a model, a working hypothesis. I deduce a possible result and verify this prediction with an experiment. I'm not saying that my prediction has the value of an experiment.

DISCUSSION OF INERTIAL AND GRAVITATIONAL MASSES

NEWTON: And you're not talking about the competing theory, neo-Newtonian mechanics. If I had established the qualitative distinction between gravitational mass and inertial mass, I would have postulated their mathematical equality. That would be the difference with my own theory. If that's all there is to the change, it can indeed be seen as an extension.

GALILEO: Bertozzi's electron acceleration experiment and the MICROSCOPE satellite experiment don't seem to discriminate in favor of one theory over the other.

KEPLER: The discrimination between the two theories would come from eclipse observations like the one made by Eddington.

GALILEO: The results of our subsequent measurements are unambiguous.

KEPLER: But why not repeat this direct measurement at every eclipse? And in the case of satellite observations, why not use Einstein's Gamma factor beforehand in data processing, which would distort the reasoning?

GALILEO: Well, speak of the wolf...

DISCUSSION OF THE SCIENTIFIC CONSENSUS

EINSTEIN, *entering*: Do you know what I've just learned?... All these hearings are directly and instantaneously communicated to the scientific community, which votes after each hearing. And do you know the results? Eight in favor of Relativity, three neutral and none in favor of neo-Newtonian mechanics. It's fair to say that Relativity enjoys a consensus in the scientific community.

GALILEO: But what is the scientific community? Who belongs to it? Do you need a diploma, co-optation by peers, allegiance to a principle or a theory? My scientific community is the members of the Tribunal of the Holy Office or the Jesuit community, I'll leave you to imagine. And there was consensus or unanimity to condemn me, wrongly in fact.

KEPLER: No, consensus is not unanimity! Because what does consensus mean in scientific terms? Consensus is agreement on something; it's a political act that enables society to move forward. Scientific progress is also based on dissensus, on disagreement. Sometimes you have to turn the table upside down, disregard old beliefs in order to move forward. I had to break the perfect image of the circle to impose the ellipse.

EINSTEIN: I agree with you. I, too, had to turn the tables when I said that time didn't seem variable, but was "really" variable. And faced with the opposition of a hundred authors, representing a certain intellectual community, I replied: "Why 100? If I was wrong, one would have sufficed".

GALILEO: But where is Noah?

EINSTEIN: He's off to join his supporters at the Café des Sciences.

KEPLER: Why not join him?

GALILEO: We've been asked to stay in this room. Taking opinions outside this committee could influence us in a partisan way.

KEPLER: You're a very obedient Catholic. What about our freedom of judgment? We have the right to be open to other sources. Listening to all these hearings on experiments alone, without devoting any time to the mathematical models behind them, seems to me to lack completeness. What are you afraid of? Do you really not want to come, Galileo?

GALILEO: We haven't heard much from you until now. But freedom of expression and movement aren't just for the Reformed! Let's go !

21. 100 AUTHORS AGAINST EINSTEIN

At the Café des Sciences counter

AN ANTI-SEMITIC BOOK?

NOÉ: Come in, my friends. Let me introduce Dr. Hanz Israel, who coordinated the book "***100 Authors Against Einstein***"[24].

GALILEO: Coincidentally, Einstein was referring to it just now.

KEPLER: Judging by your name, I'm guessing you're Jewish.

ISRAEL: I don't care about my religion in relation to this book. On the other hand, you put your finger on this paradox. Dr. Israël's book has been anathematized as an anti-Semitic opuscule. As a result, nobody reads it anymore.

GALILEO: In a way, a moral inquisition that would have blacklisted your book.

KEPLER: Why is your book described as anti-Semitic?

ISRAEL: Actually, it's not my book per se, it's a work that brings together the texts of some fifty authors and lists a good hundred opponents of the theory of Relativity.

GALILEO: Hence the name, 100 authors against Einstein.

ISRAEL: Yes, although it wasn't really against Einstein, but against his theory. In fact, his influence is such that he embodies his theory. He's even been considered the greatest genius ever since.

---

[24] https://archive.org/details/HundertAutorenGegenEinstein

GALILEO: Thank you for the rest of us who have fought to impose our ideas and conceptions.

KEPLER: That still doesn't answer my question. Why is this book called anti-Semitic, Prof. Israel?

ISRAEL: When this book came out, Einstein was asked for his opinion. His bon mot, "If I'd been wrong, then one would have sufficed", made the headlines, and public opinion stayed with it.

GALILEO: But the world of science didn't react, didn't support you?

ISRAEL: At the time, the scientific world was divided: some for, some against Relativity.

KEPLER: That still doesn't explain why this book is considered anti-Semitic.

ISRAEL: I'm coming to that. This book was published in Germany, I'm German, and the vast majority of its authors were German. However, the year after it was published, a German dictator named Adolf came to power and pursued a violently anti-Semitic policy that went so far as to exterminate the Jews.

KEPLER: Another religious war of sorts, like the one that lasted thirty years. What does that mean?

ISRAEL: More than that. Among the forty or so authors of this book, some were Jewish and suffered this repression, while others effectively rallied to this new anti-Semitic policy.

KEPLER: Just because there are a few black sheep doesn't mean the whole flock is black.

ISRAEL: Indeed. It's rather because Einstein claimed to be, and is, of Jewish descent that any statement against his theory is quickly branded anti-Semitic.

NOÉ: As far as this book is concerned, I found it a bit disparate.

ISRAEL: It's not that, each author had the freedom to express himself without a pre-established framework, hence this impression on the form. As for the content, apart from Einstein's quip, neither he nor his supporters responded to the book's critics.

KANTIAN CRITICISM

KEPLER: And what are these criticisms?

ISRAEL: Many of the authors were not pure scientists, but intellectuals, philosophers, men of universal knowledge.

GALILEO: As we were in Antiquity. Plato was a philosopher who could talk about politics and science. So could Aristotle.

ISRAEL: That's right. Our authors often referred to Kant's philosophy.

GALILEO: Who was this Kant?

ISRAEL: A Prussian philosopher who was very popular in Germany and European universities at the time. He wanted metaphysics to become an a priori science on a par with mathematics and physics. It was Newton's writings that introduced him to physics.

GALILEO: Newton will be happy to hear that.

ISRAEL: But he also quotes you, pointing out that it was because you set up experimental devices that the science of Physics was able to emerge.

GALILEO: That's flattering, thank you. But what does he say more specifically about the natural sciences, about physics?

ISRAEL: In the field of science, he wrote a theory of the heavens. He considered that the Sun was a star like any other, that stars and planets were formed from rotating clouds of gas, and he predicted the existence of galaxies, calling them "world-universes".

GALILEO: A new Giordano Bruno of sorts.

NOÉ: The future proved him right about these predictions made in the Napoleonic era. This strengthened his influence over the following centuries in this field of physics.

GALILEO: In what way?

ISRAEL: According to him, matter, energy and all our sensible experiences are necessarily inscribed in a pre-existing space and time. We can thus deduce that the curvature of Relativity's space-time caused by matter and energy is impossible.

GALILEO: How can we philosophically arrive at such a conclusion, which falls within the realm of physics and not philosophy?

ISRAEL: I can try to explain it to you in a few sentences, hoping not to betray his thought. He formulates antinomies, each consisting of a thesis and an antithesis. In our case: "*The world had a beginning in time and is limited in space*" and, on the other hand, "*The world is infinite in both time and space*". Can you follow me?

GALILEO: Thesis, antithesis... I'm waiting for the synthesis.

ISRAEL: According to Kant, no rational argument can invalidate either proposition. In other words, both can be rationally defended. Thus, according to him, there will always be an infinite time prior to any event, whether the Universe had a beginning or not.

GALILEO: That's his argument, it's one philosophical reasoning among others. How could it be more valid than that of Saint Augustine, for example, who thought that *"the Universe and time appeared simultaneously"*? In my opinion, neither religion nor philosophy should interfere with scientific theories.

NOÉ: As you well know, clerics were, and often still are, men of knowledge. In Antiquity, certain philosophers were able to free themselves from the clerics, and they called the metaphysics that dealt with ideas Superior or First Philosophy. And that which dealt with matter was called lower, second or natural philosophy. Nature being 'Phusis' in Greek, this gave rise to Physics, of which you are recognized as one of the founding fathers.

GALILEO: Whatever happens, I prefer arguments derived from or confirmed by experience. I don't believe in poets, philosophers or religious people when experience goes against them. This is the spirit in which the Chairman conducts the hearings.

ISRAEL: Kant readily admits that sensitive experience is the source of all knowledge. But he added that our reason alone enables us to perceive the world. For him, Knowledge results from the interaction between these sensory data and the cognitive structures of the mind.

KEPLER: I couldn't agree more. Without a theoretical model, we can't interpret experimental data.

GALILEO: These are philosophers' disputes that should be destroyed by the evidence of perception. In my opinion, the experimental method should combine reasoning by intuition, mathematical practice and, of course, validation by experiment.

KEPLER: Exactly what other, non-philosophical criticisms do you have of this collection?

ISRAEL: It's hard to sum up a book criticizing physical concepts in a few sentences.

THE FIRST POSTULATE

GALILEO: Let's try anyway. For example, on the first postulate, which I'm not sure now whether it's the principle of relativity that I stated.

ISRAEL: More precisely, the first postulate states that the laws of physics are invariant to changes of reference frame.

NOÉ: That's not very clear. A physical law is generally expressed by a mathematical formula. But its mathematical expression can change with a change of reference frame. For example, seen from the boat, the stone falls from the mast according to the equation of a straight line. Seen from the ground, it falls according to the equation of a parabola.

GALILEO: The relativistic formulation is indeed not my formulation. But I would agree if it meant that the law is that an object will fall at the foot of the mast, whether the ship is moving or not.

ISRAEL: Georg Wendel's criticism is rather directed against the generalization of this principle: "***The appearance of a ship traveling along the bank to be the same as if the objects on the bank were moving in the opposite direction to the direction of travel, which they cannot think of***".

GALILEO: I'm not sure I understand.

NOÉ: I think he means that when the ship is in motion relative to the shore, there's only one frame of reference where the stone falls in a straight line, and that's the ship's frame of reference. In other inertial reference frames, it falls in a parabola.

GALILEO: For me, all objects actually fall according to the natural movement of the circle.

NOÉ: Let's put it another way. In the terrestrial reference frame, only the ship moves. In other reference frames, all objects on Earth are moving.

GALILEO: So?

NOÉ: It's important to distinguish between the relativity of motion, in accordance with your principle, and the supposed relativity of reference frames. There may be one particular reference frame that is different from all the others.

GALILEO: I quite like it. All frames of reference are not necessarily equivalent. The Earth revolving around the Sun is not the same as the Sun revolving around the Earth. The Tycho-Brahé model is not the same as the Copernican model. For our solar system, there is one preferred frame of reference, and that's the Sun.

THE SECOND POSTULATE

KEPLER: And against the second postulate, on the speed of light, can you give us an example of theoretical criticism?

ISRAEL: According to Prof. Lipsius, " ***Einstein (…) turns a relative speed into a "law of nature" and thus misuses the concept of law***".

GALILEO: Speed is the ratio of distance to time. It's not a law, it's a definition. I had discussed this with Einstein on the ship, and it seemed to me that there was an incompatibility.

NOÉ: Just as it's surprising that light has a constant speed but refracts differently depending on its wavelength, as in a prism, for example.

ISRAEL: The example of the prism is also the argument of Prof. Strehl, who didn't like the fact that the refractive index of a prism varies according to the photon's wavelength.

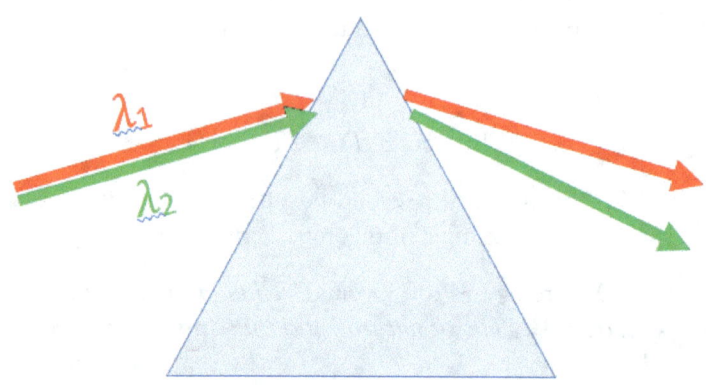

FIGURE 28: REFRACTION OF A DISC AS A FUNCTION OF WAVELENGTH

VARIABLE TIME & RELATIVE SIMULTANEITY

KEPLER: What about the corollary of constant speed, namely variable time?

ISRAEL: If we assume that the speed of light is constant and that distances are variable in Michelson's experiment, then, says Dr. Friedlaender, "**And so that the calculation is only correct, the time is put into perspective. Without any physical justification**". And on a related subject, the relativity of simultaneity...

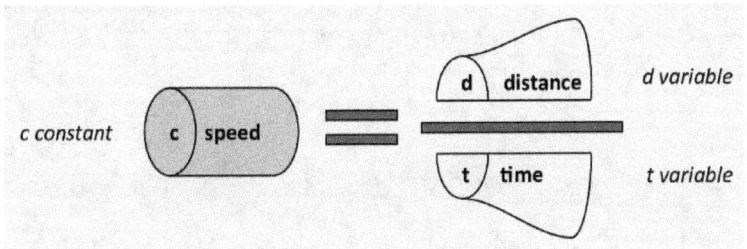

FIGURE 29: CONSTANT SPEED, VARIABLE DISTANCE AND TIME

KEPLER: Precisely how do you define the relativity of simultaneity?

ISRAEL: According to the theory of Relativity, there is no such thing as absolute simultaneity. What would be simultaneous seen from one frame of reference would not be so seen from another.

KEPLER: The concept of "relative simultaneity" is a strange one indeed - it sounds like an oxymoron. I imagine you're critical of it too.

ISRAEL: That's what Prof. Weinstein says: '*it only means that the relative simultaneity, controlled with rays, does not exist after the movement of the clock to the second clock, the absolute has remained*'.

NOÉ: It seems to me that we need to distinguish between the speed of propagation of information and simultaneity as such. Simultaneity remains absolute, whatever the frame of reference from which it is viewed. On the other hand, information cannot propagate faster than its information carrier, the photon. Just because information travels at the speed of light from one clock to the next doesn't mean that the events themselves can't be simultaneous. Two events can occur simultaneously in New York and in the depths of Siberia, but just because we know about one before the other doesn't mean they didn't happen simultaneously. This is why we need to distinguish between the speed of propagation of information and simultaneity as such.

ISRAEL: This extravagant concept of relative simultaneity often appears in relativistic demonstrations, whether for the Sagnac effect, Lorentz's equations or others.

LORENTZ EQUATIONS

KEPLER: What exactly are Lorentz equations?

ISRAEL: These are empirical formulas that form the basis of the theory of Relativity. They're fundamental. If they're wrong, the whole theory of Relativity collapses.

KEPLER: Have these empirical formulas never been proven?

ISRAEL: Yes, by Einstein himself. That's what makes him so great. But his demonstration is hardly ever taught in universities anymore. If it's correct, why isn't it taught anymore?

KEPLER: You seem very skeptical about this demonstration. I'm interested in the calculation rules. Can you tell us more about them?

ISRAEL: For example, Dr. Leroux points out that " ***Einstein did not differentiate between the fixed instantaneous values and the variable arbitrary values of t and t'*** ".

NOÉ: This is crucial, and it's a criticism that is unfortunately underdeveloped. All you have to do is indicate the variables to realize that he mixes distances traveled by different photons as well as reference frames [25][26] [27][28]. And the demonstrations of his disciples are no

---

[25] **Criticism of Einstein's demonstration**
[26] https://www.youtube.com/watch?v=koPnW0mXcvI
[27] https://www.youtube.com/watch?v=lfgyexpq_BQ
[28] https://www.youtube.com/watch?v=E8omcJ9fuYM ;
https://www.youtube.com/watch?v=41qlajZth0Y ;
https://www.youtube.com/watch?v=IJYo31BAxOA ;
https://www.youtube.com/watch?v=SMiAAwHPAsM ;
https://www.youtube.com/watch?v=piRlSVPdwkE ;
https://www.youtube.com/watch?v=WL-l-8D4ce8

better[29]. But who today would dare to seriously criticize the demonstration of the super-genius Einstein?

GALILEO: Clearly you, at the risk of being seen as a fool, a megalomaniac or at best presumptuous, unless posterity proves you right, as it did in my case.

KEPLER: I'll take time to examine your criticisms of relativistic calculations.

GENERAL RELATIVITY

GALILEO: But tell me, Dr. Israel, have you made any criticism of the second part of this theory, namely General Relativity?

ISRAEL: Yes, of course. I personally, in all modesty, pointed out that '***The gravitational field cannot be replaced by an accelerated system, since both systems are not equivalent***'. This is the same idea defended by Prof. Lipsius: '***Inertia and gravity can only be exchanged as long as homogeneous gravitational fields are taken into account. But an absolutely homogeneous gravitational field is a mere thought. . . .***'.

GALILEO: Don't you have an example, so that I can understand better?

NOÉ: A gravitational field is radiant and decreases with the square of the distance.

GALILEO: That it's radiant is obvious. That it decreases with the square of the distance, I can believe. But what does that imply?

NOÉ: This has two consequences. The first, undeniable, is that the gravitational acceleration lines are not parallel, since they radiate out from the center. The second is that the attraction away from the

---

[29] **"Reply to "A Simple Derivation of the Lorentz Transformation"**, https://www.scirp.org/journal/paperinformation.aspx?paperid=81181

object is weaker than near the object. This can be represented by these concentric dotted circles. The closer these circles are to each other, the stronger the attraction.

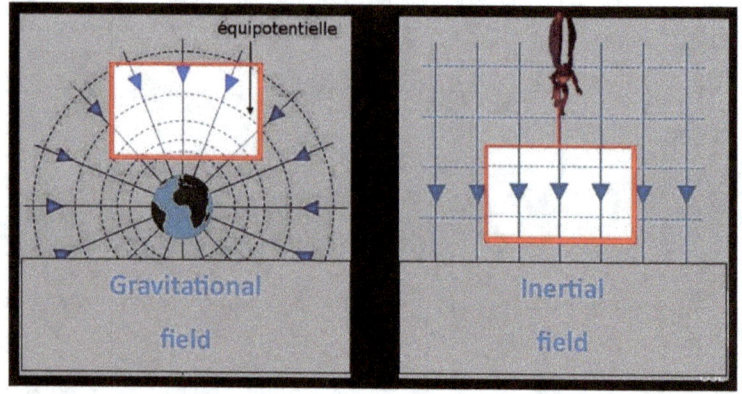

FIGURES 30: GRAVITATIONAL FIELD & INERTIAL FIELD

GALILEO: This diagram [on the left] seems clear.

NOÉ: This diagram is different for an inertial representation. Imagine that you are enclosed in the white box and that a pixie is pulling the box with uniform acceleration.

GALILEO: As you know, I also enjoy thought experiments.

NOÉ: Inertial acceleration lines are parallel to each other in the direction of motion.

GALILEO: Of course.

NOÉ: And the inertial acceleration is the same throughout the box. This can be represented by horizontal lines all at the same distance from each other, since there is no place where the inertial acceleration is stronger than at another point.

GALILEO: Certainly.

NOÉ: Then you realize the differences. Gravitational acceleration is radiant, inertial acceleration is parallel to itself. The force of gravitational acceleration is stronger near the attractive mass than far from it, while the force of inertial acceleration is uniform, the same at every point of the box. How can we conclude that a gravitational field is equivalent to an inertial field?

KEPLER: Precisely, it would seem that the scientists of the time accepted this despite the fact that they had to be aware of this pattern.

NOÉ: That makes me wonder. Relativists call it the equivalence principle, which stipulates, as we often forget, that it's only true locally, in a small space and over a short period of time. This shows, on the one hand, that it's not true in general. And secondly, even locally, if you increase the precision of the measurement, it's not true. This non-questioning really makes me wonder, especially as it is the basic principle of the theory of General Relativity.

KEPLER: I suggest we sit down for a moment. Now that we've discussed this theory of Relativity, I'd like to ask you a few questions about the neo-Newtonian theory.

NOÉ: With pleasure.

GALILEO, WAKE UP!

22. ASYMPTOTIC SPEED S

At a table in the Café des Sciences

KEPLER: I have a question. According to the theory of Relativity, nothing can go faster than light. Bertozzi's experiment with electrons seems to confirm this. Fizeau's experiment with moving water shows that Newtonian addition of velocities is false and that relativistic addition of velocities is true, perhaps conceding to you that there are some measurement uncertainties on the velocity of water. The law of relativistic velocity addition prevents us from going faster than the speed of light. What happens on your side, according to the theory of neo-Newtonian mechanics?

NOÉ: As you said, nothing seems to go faster than the speed of light, apart from a few measurement uncertainties. According to the theory of Relativity, no uncertainty is allowed, no other case is possible. This speed of light is called "c", from the Latin "celeritas", meaning speed. In English, this has given rise to the word "celerity". According to the theory of Relativity, this speed "c" is invariable.

KEPLER: We know that now. But, according to neo-Newtonian mechanics, can we exceed the speed of light?

NOÉ: I'm coming to that. While the speed of light "c" is invariant in Relativity theory, it is very slightly variable in neo-Newtonian mechanics, and I'll explain why later. And **according to neo-Newtonian mechanics, there's also a threshold speed, called "s", which comes from "speed". This is more of an aSymptote or** limit than an actual speed. It is defined from an electromagnetic point of view[30], and this is not a property, by :

$$s = \frac{1}{\sqrt{\varepsilon_0 \mu_0}}$$

with μ_0 magnetic permeability of vacuum and ε_0 dielectric permittivity of vacuum.

KEPLER: Isn't this formula also valid in Relativity?

NOÉ: Yes, it is, except that it's a property of Relativity theory and a definition of neo-Newtonian mechanics.

KEPLER: Is that why you didn't keep the "c" symbol?

NOÉ: That's right. It's to distinguish the speed of light, "c", which is physically reached by photons and which is slightly variable, from this asymptotic speed "s", unattainable even by photons.

KEPLER: How much do they differ from each other?

NOÉ: Expressed in meters per second, infinitesimally, perhaps only 0.00002 Å/s! But we're at the limit, practically sticking to the asymptote. To increase our speed by a few billionths of a metre per second, without exceeding this speed limit, we need to deploy phenomenal amounts of energy, as Bertozzi's experiment with accelerated electrons shows.

KEPLER: I understand that the distinction between Relativity and neo-Newtonian mechanics is about the nature of the photon velocity, but not about the numerical values, which are in practice identical. So, in practical terms, what difference does it make if it's the same limiting value?

---

[30] "Let's free Newtonian Mechanics from the 'principle' of equivalence!", https://www.gsjournal.net/Science-Journals/Research%20Papers-Relativity%20Theory/Download/7499

NOÉ: According to the theory of Relativity, the speed of a photon is invariant and equal to "c" for any observer, whatever the speed and direction of his reference frame, and whatever the speed of the observer taking the measurement.

KEPLER: It's true that this relativistic concept is not intuitive.

NOÉ: According to neo-Newtonian mechanics, the speed of light can be variable, but I need to introduce another concept, the two velocity additions, are you ready?

KEPLER: Yes, of course, I'm listening!

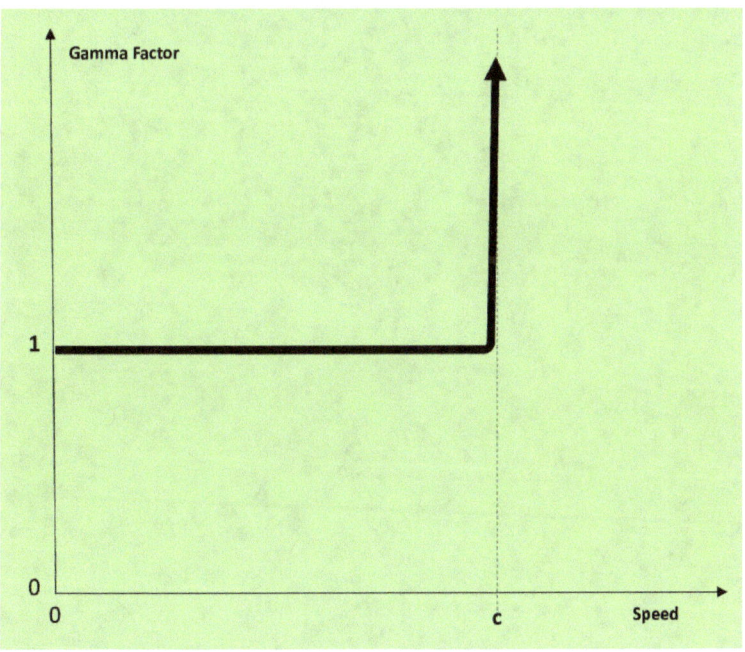

FIGURE 31: GAMMA FACTOR AS A FUNCTION OF SPEED

GALILEO, WAKE UP!

23. THE TWO SPEED ADDITIONS

NOÉ: In Relativity theory, and in Newtonian mechanics too for that matter, we don't distinguish the speed of an object coming towards us from our own speed going towards the object.

KEPLER: This comes from the Galilean principle of relativity.

NOÉ: We saw with Galileo that it's not fundamentally the same thing to observe from the shore and to observe from a moving boat. In the first case, nothing moves except the boat. In the second case, everything moves: the shoreline, mountains, other floating objects, etc. Certain reference frames are special.

KEPLER: What are you getting at?

NOÉ: There are two laws of addition of velocities, that of objects and that of reference frames.

KEPLER: It's a new concept! You're arousing my curiosity.

NOÉ: Let's remember that speed is not a physical datum in the same way as gravitational mass, length or duration. Speed is the ratio of a distance traveled in a reference frame to a duration.

KEPLER: Yes, of course.

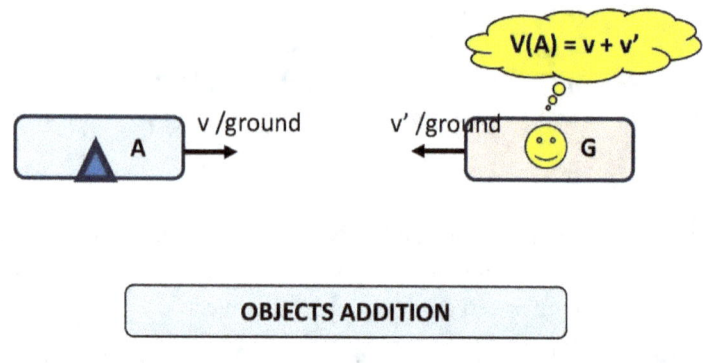

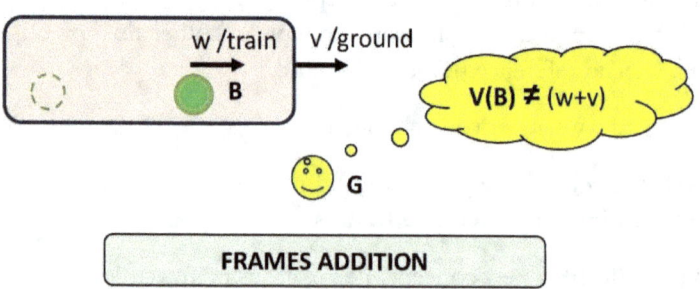

FIGURES 32: <u>THE TWO TYPES OF SPEED ADDITION</u>

NOÉ: Let's look at the first, simplest case, the addition of object speeds. Alice is seated on a train which is moving forward in relation to the platform at speed v. Galileo is seated on another train moving in the opposite direction, at speed v'. How fast does Galileo see Alice moving towards him?

KEPLER: It's obvious, at the sum of speeds v+v'.

NOÉ : Indeed, the addition of the velocities of objects is also what we call the Newtonian addition of velocities.

KEPLER: So what is the sum of the velocities of the reference frames?

NOÉ: Albert is now moving in the train at speed w relative to the train. Galileo, our observer, now on the platform, sees the train moving at speed v. How fast does Galileo see Albert moving?

KEPLER: It's obvious, at the sum of speeds v+w.

NOÉ: This seems incontestable at low speeds, but it's not true at ultra-high speeds. In fact, this is what the theory of Relativity expresses. If Albert is a photon of light, Galileo will see him approaching at speed c, not at speed v+c.

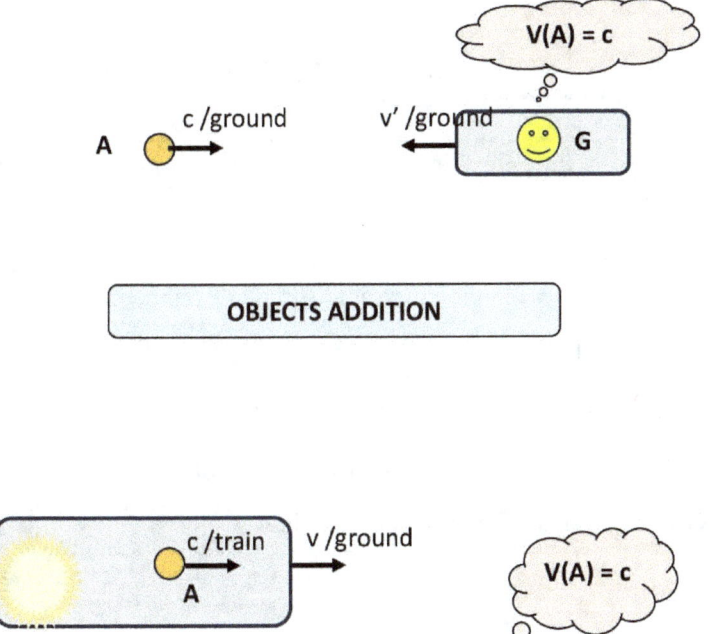

FIGURE 33: VELOCITY ADDITIONS ACCORDING TO THE THEORY OF <u>RELATIVITY</u>

KEPLER: This result seems to be in line with the theory of Relativity. Would you agree now?

NOÉ: In this second case, yes, but with only a tiny difference in speed. We saw in Bertozzi's experiment that, even if ultra-fast electrons are given more energy, they do not exceed a threshold speed. So, in neo-Newtonian mechanics, in the case of a photon emitted at speed c on a train, its speed c', as seen from the platform, will remain below the threshold speed s.

KEPLER: How is this possible?

NOÉ: **The extra energy is essentially transformed into inert mass. In this case, the result is a photon:**

> **Addition of the velocities of the reference frames: $c' < s$ always**

KEPLER: So, according to you, nothing can go faster than light or, as you say, faster than the threshold speed "s", right?

NOÉ: Absolutely not!

KEPLER: I thought I understood you, but you've lost me.

NOÉ: You've forgotten the first case, the addition of objects.

KEPLER: What difference does it make?

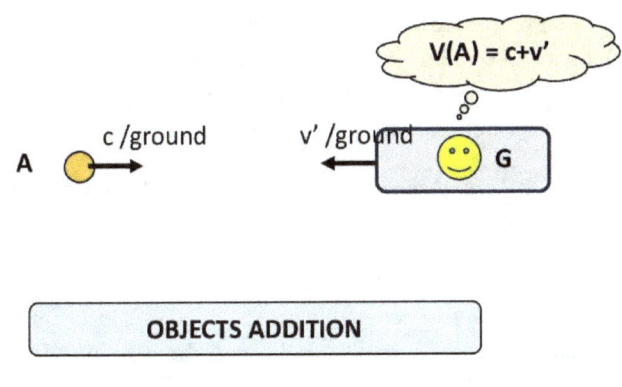

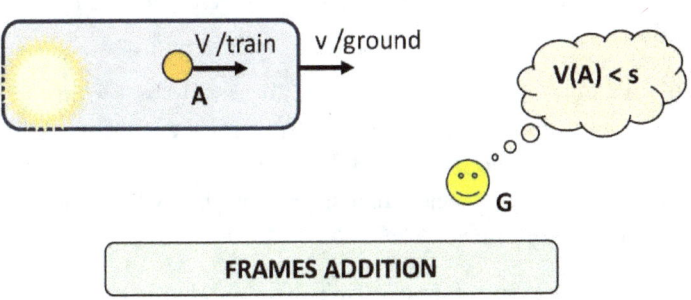

FIGURES 34: SPEED ADDITIONS ACCORDING TO <u>NEO-NEWTONIAN</u> MECHANICS

NOÉ: Let's assume that the train Alice is on is travelling at speed "c" - in other words, Alice is a photon. But, unlike the previous case of a change of reference frame, where there was an input of energy that was transformed into inert mass, the situation is unchanged. Galileo will see Alice arrive at speed c+v'.

Addition of object speeds : $c + V > s$ possible

In this case, the speed of the photon as seen by Galileo is greater than the speed "c". According to neo-Newtonian mechanics, a photon is emitted at speed "c" as seen by an observer located in the frame of reference of the photon emitter. For another observer moving at speed v relative to the photon emitter's frame of reference, he or she will see this photon moving at speed "c-v" or "c+v" depending on the direction taken by this other observer.

KEPLER: This dual concept of adding speeds, objects and reference frames, is interesting...

NOÉ: Thank you. This comes from speed, which is not a physical datum like mass or length, because speed is the result of a ratio, a calculation. This is why, depending on the case, one formula or the other must be used to add up the speeds.

KEPLER: I'd like to come back to the addition of the velocities of the reference frames. At the Fizeau hearing, you presented two different formulas for adding the velocities of reference frames. Why did you do this?

NOÉ: It's possible to find Fizeau's result exactly by starting from kinetic energy, but I confess my preference for the velocity addition formula based on momentum. But it's Fizeau's experiment, which will have to be more precise than at present, that will decide which formula is the right one.

KEPLER: Apart from Fizeau's, isn't there another experiment that could help us decide?

NOÉ: I could see another one, but it's not an experiment, it's an astronomical observation that could go in the direction of momentum.

KEPLER: An astronomical observation, you say, you interest me. Please continue.

24. ELLIPSE *vs.* OVOID

At the same Café table

KEPLER: What astronomical observation would you like to talk to me about?

NOÉ: You've worked extensively on the orbit of Mars.

KEPLER: Absolutely, and I must say successfully. Mars, the red planet, made my blood boil. After years of studying it, I was finally able to determine that its trajectory was elliptical.

NOÉ: You've also worked on Mercury's orbit.

KEPLER: Yes, but I admit with less success. Its orbit is clearly not circular, but my model of the ellipse doesn't exactly match its trajectory either. There are some irregularities, some small differences with this model of the pure ellipse.

NOÉ: These irregularities were largely explained by Isaac Newton who, on the one hand, established that the trajectory was indeed an ellipse...

KEPLER: I'm delighted that he's taken up my ellipse model. But where do these irregularities come from?

NOÉ: And, secondly, that this elliptical orbit gradually rotated around the Sun, mainly under the influence of Venus and Jupiter. This is known as the precession of Mercury's perihelion, i.e. a gradual change in the orientation of Mercury's orbit.

KEPLER: So that's the end of the matter!

NOÉ: For the most part, yes. But the devil is sometimes in the detail. While Newtonian mechanics explains more than 90% of this precessional motion of Mercury's perihelion, there's still a small 10% to explain. This is known as the Mercury perihelion anomaly.

KEPLER: Couldn't this be due to some uncertainties in the observation measurements?

NOÉ: Maybe only marginally, but the difference is real.

KEPLER: What are you getting at?

NOÉ: The theory of Relativity fully explains this discrepancy, based on the fact that Mercury is the closest planet to the Sun, where the distortion of space-time is greatest. This observation is often put forward as proof of this theory.

KEPLER: I understand your embarrassment.

NOÉ: Absolutely not! During our discussion on "100 authors against Einstein", I explained to you some simplifications that seem to me to be mathematically abusive in the demonstration of the Lorentz equations. Similarly, two simplifications in the relativistic demonstration of Mercury's perihelion anomaly seem to me to be equally abusive.

KEPLER: Decidedly! I'll be watching your criticisms of this relativistic calculation carefully. And, by the way, does your theory explain this anomaly?

NOÉ : Indeed, on the fact that Mercury is the fastest planet in the solar system, hence its name since antiquity, but leading to a solution you may not like.

KEPLER: What do you mean?

NOÉ: **According to neo-Newtonian mechanics, orbits are not elliptical trajectories, but ovoid, egg-shaped trajectories**[31] !

---

[31] **"About the Ovoid Orbits in general, and Perihelion Precession of Mercury in particular,"** http://www.mrelativity.net/Papers/51/Mercury%20Millennium%20Serret%205%20janvier%202018.pdf

KEPLER: Don't get me wrong, I had the same intuition. I started with the ovoid before moving on to the ellipse.

NOÉ: There are an infinite number of ovoid models. You took one, the egg model you'd created mathematically, it didn't correspond to the orbit of Mars, and you switched to the ellipse. Now, the eccentricity of orbits, which corresponds to the flattening out with respect to a perfect circular trajectory, is low for the planets in our solar system. Elliptical models therefore differ little from those of an egg. The conceptual problem with the ellipse model is that, while the Sun is located at the first focus, at the second focus there is physically nothing. It's a bit like Ptolemy's system: for the model to be consistent with observations, it's necessary to invent a fictitious point, the second focus, around which the star would revolve...

KEPLER: I'm well aware of this, which is why I started with an ovoid model, with a single focus where the sun was located. In their revolution, each planet moved towards or away from the Sun, depending on magnetic forces that I didn't know how to evaluate. Do your model and your calculations correspond to the observations of all the planets?

NOÉ: The orbits of the planets in our solar system have low eccentricity, with little difference between the elliptical and ovoid models. The greatest eccentricity is found in the orbit of Mercury, the closest and fastest planet to the Sun.

KEPLER: In a way, if I may say so, I've gone from the ovoid to the ellipse, and you've gone back from the ellipse to the ovoid.

NOÉ: That little bit of humor might make you think I'm backtracking. No, because I don't end up with the same ovoid model as you.

KEPLER: No offence. Apart from Mercury's trajectory, do you have any other examples that would support your model?

NOÉ: We're currently measuring the orbit of the star S2 at the center of our Galaxy. There could also be the orbit of certain comets, objects made up of rock and ice that move through our solar system.

KEPLER: Comets, those hairy stars according to their Greek etymology. I had personally observed two of them, but hadn't managed to describe their trajectory.

NOÉ: Because we only see a portion of their path. Even today, we find it difficult to model the trajectories of comets with eccentricities of up to one unit.

KEPLER: If their eccentricity equals or exceeds unity, then it's a parabolic or hyperbolic motion, i.e. trajectories that go out into infinite space. These comets will not return.

NOÉ: What's surprising is the number of parabolic trajectories, i.e. with eccentricity exactly equal to one. If one of these comets were to return in the future, it would mean that the mathematical model applied to their trajectory was not correct.

KEPLER: Only time will tell. In the meantime, what happened to Galileo?

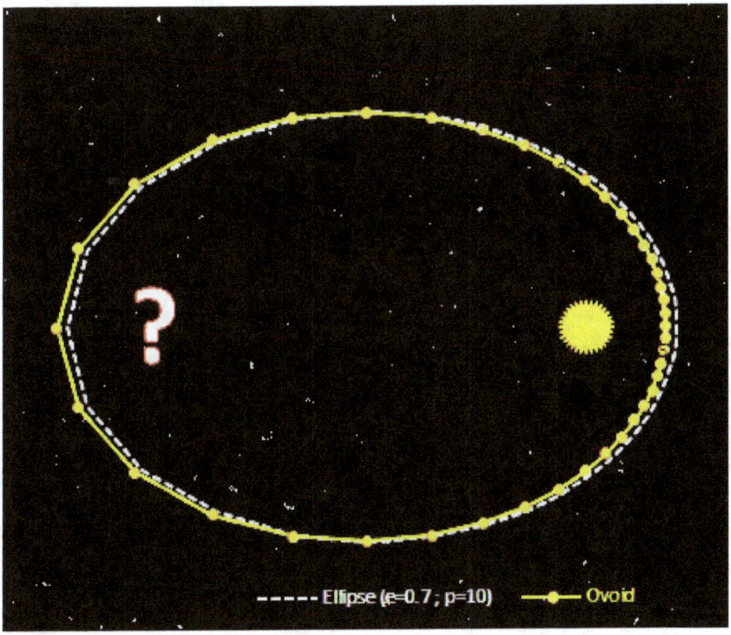

FIGURE 35: ELLIPSE VS. OVOID

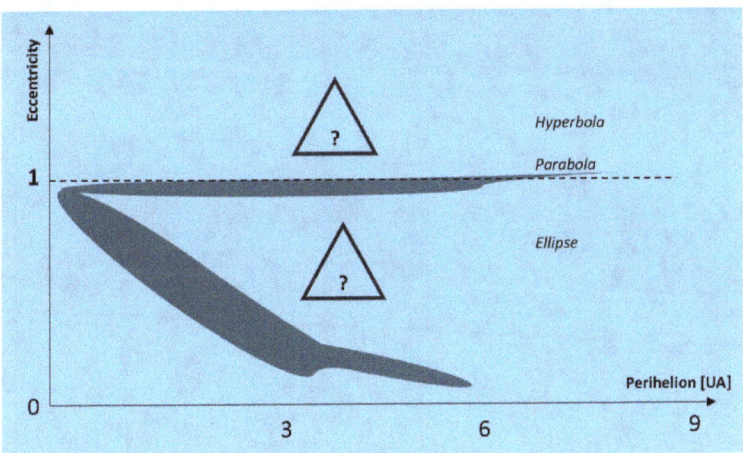

FIGURE 36: ELLIPTICAL ECCENTRICITY OF COMETS AS A FUNCTION OF THEIR PERIHELION

GALILEO, WAKE UP!

25. THE PHOTON

NOÉ: We lost you, where were you?

GALILEO: In charming company, with someone called Susan Jocelyn, who has made a remarkable discovery.

NOÉ: You're incorrigible! And what were you discussing, if not indiscreetly?

GALILEO: Not at all! She taught me about a mysterious cosmic object called a pulsar.

NOÉ: What did she tell you about it?

GALILEO: This is an astronomical body she helped discover as a student.

NOÉ: That's why it's said that she was unfairly denied the Nobel Prize. But what did she tell you about the pulsar?

GALILEO: This is the residue of an ancient star, particularly small and dense, which has the peculiarity of rotating very rapidly on itself and emitting beams of light and other types of radiation into space. It's a bit like a cosmic lighthouse, spinning and emitting flashes of light at regular intervals.

NOÉ: That's a good summary. Did she tell you about the Measure of Dispersion, DM in English?

GALILEO: Of course, young man! The Measure of Dispersion means that the lowest frequencies arrive a little later, in dispersed order if you like, than the highest frequencies.

NOÉ: And what's his explanation?

GALILEO: Better ask him.

NOÉ: Regarding the staggered arrival of waves according to their frequency, this is explained by slightly different propagation speeds.

GALILEO: But wouldn't that contradict the constant speed of light in Relativity?

NOÉ: I think so.

GALILEO: I meant, how do relativists explain this?

NOÉ: Relativists postulate that light travels at a constant speed in a vacuum. But they consider that the cosmos is not empty, that it is filled with various particles and gravitational effects that affect the propagation of the photon as a function of its wavelength through the ultimately non-empty medium of interstellar space.

GALILEO: The constitution of interstellar space is a bit beyond me. How do you explain it?

NOÉ: Neo-Newtonian mechanics explains this variation in photon speed not by the photon's environment, but by the photon itself. Let me explain. The higher the photon's frequency, the faster it moves. Since frequency is proportional to energy, this means that the more energy a photon has, the faster it moves, and therefore the faster it arrives.

GALILEO: This seems to make a certain amount of sense.

NOÉ: Thank you. But this is not the view of relativists, who postulate that the photon can only propagate at speed "c" in a vacuum, because this particle would have the particularity of having no mass.

GALILEO: A particle without mass, you say, I don't quite understand.

NOÉ: According to the theory of Relativity, the photon is a massless particle of energy, because no mass can reach the speed of light. And since it travels at the speed of light, time would no longer pass in the photon's frame of reference, and it would always be the same age.

GALILEO: You've lost me! The photon is a massless, ageless particle.

NOÉ: It's quite incomprehensible indeed. A photon that has travelled through space for billions of years would not be subject to the passage of time, but when it arrives in water, its speed slows down, that's an established fact, and so time would have a hold on it.

GALILEO: You can tell me all you like about possible paradoxes of Relativity, but I imagine that relativists won't agree with you.

NOÉ: Maybe they wouldn't agree, but the main thing is that they don't talk about that consequence.

GALILEO: A gap, no doubt. Tell me more about your conception of the photon.

NOÉ: As you may have guessed, **according to neo-Newtonian mechanics, the photon has mass - extremely low mass, but mass nonetheless**[32]. A similar case occurred with the neutrino, which was once considered a massless particle, before it was finally established that it did have a mass, albeit a very small one. The mass of the photon would be ten billion times lower than that of the neutrino, according to my estimate.

$$m_{photon} \approx 10^{-47} \, kg$$

Galileo: With such a low value, it can be considered massless in practice.

---

[32] **"The Mass of a Photon estimated from the Pulsar Dispersion Measurement (DM)"**, https://www.gsjournal.net/Science-Journals/Research%20Papers-Relativity%20Theory/Download/7490

Noé: From a practical point of view, yes, but not from a theoretical point of view. We're talking theory here. But isn't that Newton coming?

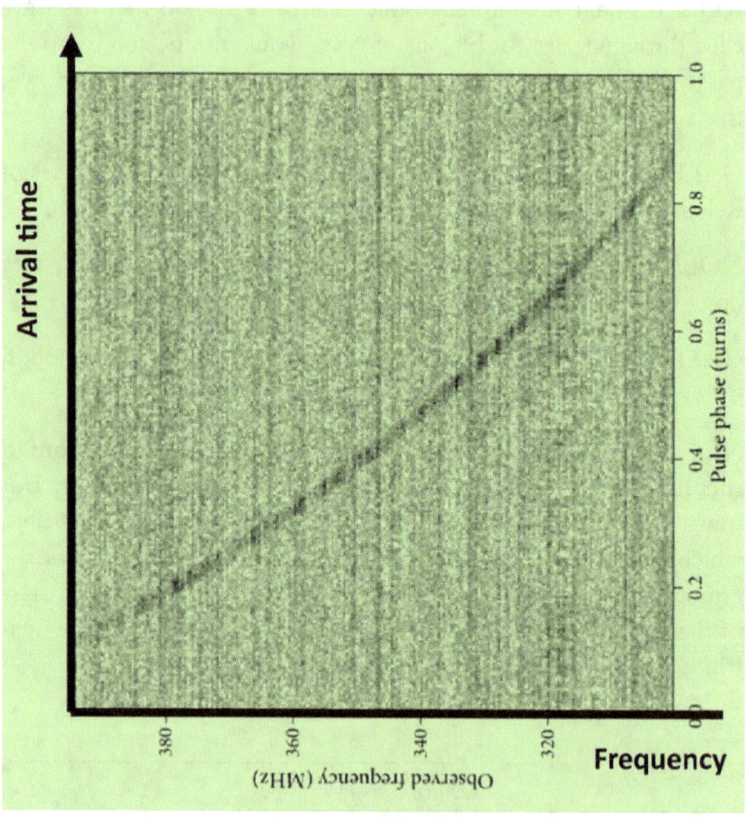

FIGURE 37: MEASURING PULSAR DISPERSION

26. ENERGY

NEWTON: I've been looking for you. We've been notified that the auditions won't resume until tomorrow, and they'll be on cosmology. May I join your discussion?

NOÉ: We're happy to talk about the photon particle and its mass.

NEWTON: As far as I know now, it's thanks to Einstein that the hypothesis of light made up of particles has come back into favor with scientists, as I thought it would, rather than considering light as a wave.

NOÉ: Yes, but with some subtleties. According to the theory of Relativity, photons are massless particles, but sensitive to gravitation! Understand who can.

GALILEO: We've just seen that the photon could have mass, according to neo-Newtonian mechanics.

NEWTON: Isn't there also a story of equivalence between energy and mass, which could explain this gravitational deviation of the photon?

NOÉ: Indeed. You're referring to the famous formula $E=mc^2$. Depending on the photon's energy, this would give it a variable mass. This formula linking mass to energy is associated with Einstein, but it's not strictly speaking relativistic; it doesn't involve variable time.

NEWTON: Do you then take it over?

NOÉ: Yes, in a sense. **A similar formula can be demonstrated in the neo-Newtonian framework using only mathematical properties**[33] :

$$\boxed{E = (\gamma . m_g). s^2}$$

with E energy, γ Lorentz factor, m_g gravitational mass and s asymptotic velocity

KEPLER: Except that, in Relativity, if despite everything the photon has no mass, is pure energy, this would mean that even energy would be deflected by gravitation. All forms of energy would be deflected by gravity. Relativity remains difficult to grasp.

NEWTON: In any case, Relativity rehabilitates the corpuscular hypothesis of light that I had put forward in the face of the undulatory hypothesis.

NOÉ: It's not quite as clear-cut as that. In Relativity theory, we speak of wave-particle duality, i.e. light behaves both as a particle and as a wave.

NEWTON: That's not a clear-cut opinion. It shows a lack of understanding of the matter, as the two concepts are incompatible. What do you propose to do about this?

NOÉ: **Perhaps a particle that moves by spinning,** that's a theoretical avenue that could be explored.

NEWTON: This is a direction you're proposing, but haven't explored.

NOÉ : Not really.

---

[33] **"How to Demonstrate the Lorentz Factor: Variable Time vs. Variable Inertial Mass",** https://www.scirp.org/journal/paperinformation.aspx?paperid=54203

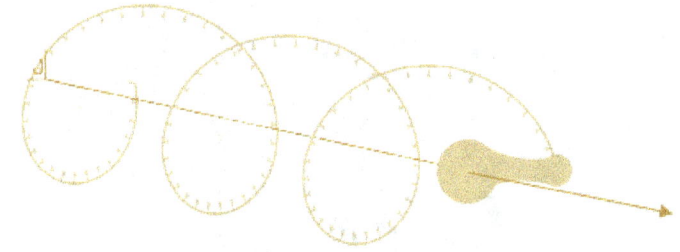

FIGURE 38: HYPOTHESIS ON PHOTON DISPLACEMENT?

GALILEO, WAKE UP!

27. THE STRENGTH

NEWTON: There's one subject that particularly interests me, and that's force. What's your analysis of it?

NOÉ: You've made a distinction between inert and serious mass. I would add that we can also distinguish between mass that attracts and mass that is attracted. Attracting mass is gravitational mass and attracted mass is inert mass. As they are different in nature, they can a priori be different in value.

NEWTON: I grant you, it was experience that led to their mathematical equality.

NOÉ: This is the experiment at normal speeds. At ultra-high speeds, speeds close to the speed of light, this no longer works precisely. I mentioned this during the Bertozzi hearing, and we've just discussed it with Kepler for Mercury, the fastest planet in our solar system. These experiments and observations can all be interpreted according to neo-Newtonian mechanics.

NEWTON: Does this distinction between gravitational mass and inert mass imply any adaptations in the formulas?

NOÉ: In your law of universal attraction, **it is indeed necessary to introduce the Gamma factor to make it clear that one mass is attractive and the other mass is attracted**[34].

$$F_{gravitationnelle} = G \cdot \frac{M_g \cdot (\gamma \cdot m_g)}{r^2}$$

---

[34] "**About the Ovoid Orbits in general, and Perihelion Precession of Mercury in particular,**" http://www.mrelativity.net/Papers/51/Mercury%20Millennium%20Serret%205%20janvier%202018.pdf

With M_g attractive mass and m_g attracted mass

But the most modified formula is the expression of force as a function of inert mass.

NEWTON: Do you mean to say that you also add the factor γ to the acceleration formula to make it clear that the force is exerted on inert mass?

NOÉ: Better than that. You defined force as being capable of modifying momentum. The quantity of motion is the major physical concept; force is merely its derivative with respect to time.

NEWTON: Absolutely.

NOÉ: Thus, **the expression for inertial force becomes**[35] :

$$F_{inertial} = \gamma^3 . m_g . a$$

with F for inertial force, mg the gravitational mass, a the acceleration and γ the Lorentz coefficient.

NEWTON: This power of 3 in the formula may seem strange at first, but it's well seen with the derivation from momentum! Do you have an experiment that would corroborate this formula?

NOÉ: Yes, with the expression of scattered energy in a synchrotron. A synchrotron is a particle gas pedal designed to generate high-energy electromagnetic radiation. The expression of energy lost 'calculated' according to neo-Newtonian mechanics is homogeneous with the expression **of energy lost 'measured' by radiation in a synchrotron**[36] .

---

[35] **"Net Force F = γ³ ma at High Velocity"**, https://www.scirp.org/journal/paperinformation.aspx?paperid=66042

[36] **"Net Force F = γ³ ma at High Velocity"**, https://www.scirp.org/journal/paperinformation.aspx?paperid=66042

$$W' = \gamma^3 \left(\frac{\gamma V^2}{r}\right).V.m_g.t$$

NEWTON: Frankly, this is getting more complex, and deserves thought and analysis. It's late now, so let's get some rest before tomorrow's discussion of cosmology.

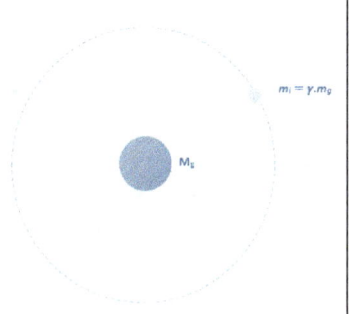

FIGURE 39: ATTRACTIVE GRAVITATIONAL MASS AND ATTRACTED INERT MASS

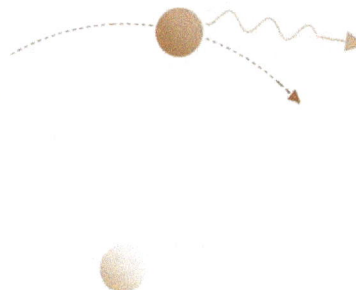

FIGURE 40: SYNCHROTRON RADIATION

GALILEO, WAKE UP!

DAY THREE (COSMOLOGICAL)

From the Hubble Constant to the Cosmic Microwave Background

GALILEO, WAKE UP!

28. HUBBLE'S CONSTANT (?)

THE PRESIDENT: Before resuming the hearings, a word about the previous hearings. We have heard testimony from experts, experimenters and observers of physical phenomena contributing to the foundations and practical applications of each of the two theories. We focused on the facts, knowing that they only make sense if they are interpreted. On the other hand, following yesterday's interruption of the meeting, I would like to reiterate that the aim of this commission is not to enter into a debate of ideas on the respective theories, a debate which will remain sterile, with everyone sticking to their positions. The aim of this commission is to examine the experiences and observations in favor of one theory or the other. Having set the scene for these hearings, we will now move on to an area of interest to a wider audience: cosmology. Of course, this will no longer involve laboratory experiments, or even manipulations in which parameters can be changed. It will be a matter of simple observations. They can be interpreted according to the reading grid of each of the two theories involved. Compared to the two previous days, you can now, if need be, use your own theoretical framework, on the express condition that you keep your presentation brief and to the point.

That said, before we hear the next testimony on cosmology and observations of phenomena at distances, I'd like us to agree on the vocabulary we'll be using here to apprehend distances in cosmology:

- Nearby distances are those within our own Galaxy. They can be determined by stellar parallaxes, i.e. observing the apparent variation in a star's position when viewed from two different points on Earth six months apart.
- Average distances are those of galaxies in relative proximity to our own. These can be determined using "standard candles" such as Cepheid variable stars. These are standard

candles in the sense that, by hypothesis, each Cepheid emits the same luminous intensity; by comparing their observed luminosity with their intrinsic or emission luminosity, we can determine their distance.

- Distant galaxies are those that can only be seen with standard candles, such as the rarer Type Ia Supernovae.
- Ultra-far distances are those of galaxies that can only be estimated by redshift, using Hubble's law, which we will now describe in detail.

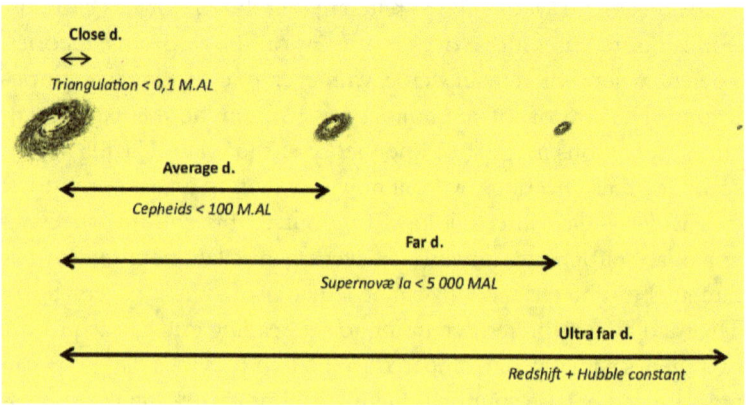

FIGURE 41: CONVENTION ON COSMOLOGICAL DISTANCES

With that introduction, please introduce yourself.

HUBBLE: Thank you, Madam President, and good morning, gentlemen. My name is Edwin Hubble, astronomer and, incidentally, Master of Jurisprudence. As you just reminded us, distances within our Galaxy are measured by stellar parallax. It's the only absolutely certain measurement, based on geometry.

EINSTEIN: Simple precision, based on Euclidean geometry, provided our Universe is flat. I apologize for the interruption.

HUBBLE: Please, Professor Einstein. This is indeed a measurement made under the assumption that Euclidean geometry is valid in our Universe, Professor Einstein having just reminded us that we can make another assumption, namely that we would be in a curved space-time.

THE PRESIDENT: That's why I said that medium and large distances are apprehended and estimated, not determined. Indeed, measurements based on luminosity and redshift are also based on the assumptions that Supernovae and Cepheids have the same intrinsic luminosity throughout the Universe, and that the redshift assumption is correct. What is your observation, Mr. Hubble?

HUBBLE: In particular, using the Mount Wilson telescope in California, I observed what were thought to be nebulae in our Galaxy, which turned out to be other galaxies around our Galaxy. Then, using the "standard candle" method applied to Cepheids, I was able to determine the distance to these galaxies.

THE PRESIDENT: And you were radically wrong! It wasn't an error of a few percent in measurement uncertainty, but the order of magnitude that was wrong. You overestimated the distances by a factor of seven, estimating these galaxies to be seven times closer than they turned out to be.

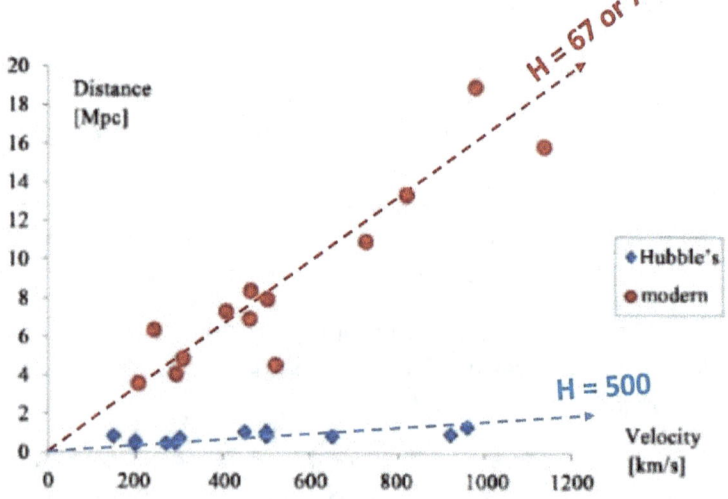

FIGURE 42: INITIAL (500) AND CURRENT (67 OR 72?) HUBBLE CONSTANT VALUES

HUBBLE: It's true that the Cepheid method was new and needed to be clarified, particularly as regards the absolute magnitude of these Cepheids. The distance correction was made some ten years later. But the principle remains correct.

THE PRESIDENT: Everyone can make mistakes, that's not what I wanted to say, I just wanted to emphasize that galaxy distances are apprehended or estimated. Please continue.

HUBBLE: Then I analyzed the light spectra of these galaxies. I observed that the spectral lines generally show a red shift. This is characteristic of a source that is moving away.

THE PRESIDENT: You've determined that all the galaxies are moving away from our Galaxy, and that's already a crucial point. How fast?

HUBBLE: To determine the speed at which galaxies are moving away, I used Doppler's relativistic law of light rays.

THE PRESIDENT: For our audience, let's recall that Doppler-Fizeau's law describes the change in frequency perceived by an observer when the wave source and the observer move relative to each other. A well-known application is the speed of a racing car coming towards you and then moving away, its sound changing from high to low. Continue.

HUBBLE: The Doppler-Fizeau effect is valid for sound moving through still air. Since the photon does not move in an immobile ether, it is no longer possible to use this Doppler-Fizeau law. This is why we have to use Doppler's relativistic law of light-wave propagation to determine the speed at which galaxies are moving away from us. These are not the same laws.

EINSTEIN: Since the Relativistic Doppler Effect, or RDE, always includes the word Doppler, there's a risk of confusion with the Doppler-Fizeau effect. This is why we prefer to speak of the expansion of the Universe, except in the case of the Diffuse Cosmological Background.

NOÉ: Clearly, the theory of Relativity is fundamentally incompatible with the Doppler-Fizeau effect, except when it suits you.

THE PRESIDENT: You'll have a chance to explain your point of view later, Mr Noé. Please continue, Mr. Hubble.

HUBBLE: By measuring, or estimating, the speed of galaxies, I discovered something extraordinary, which from my point of view, albeit only as an astronomer, should have won me the Nobel Prize in Physics.

THE PRESIDENT: Please, let's get to the point.

HUBBLE: Galaxies move away from each other at a speed proportional to their distance!

THE PRESIDENT: Why is that so extraordinary?

HUBBLE: At the time, people thought, and I thought, that the Universe was stable, that it was like that and had always been like that. With galaxies actually moving away from each other, some people ran the film backwards. Going back in time, they have imagined that galaxies are moving closer together. In other words, all the matter in the Universe was originally concentrated in a single point. To have the Universe as we know it today, there must have been an initial explosion, which was called the Big Bang!

THE PRESIDENT: Deduction is fine, but I'd like to come back to your observations and their immediate results. You haven't quite answered the question: how fast are these galaxies moving away?

HUBBLE: At speeds proportional to their distances, as I was saying, and therefore according to a constant element, now called the Hubble constant, which I had initially estimated at 500. This invalidated the hypothesis that the Big Bang took place two billion years ago.

THE PRESIDENT: That is, after the creation of the Earth...

HUBBLE: Hence the problem! But, as you pointed out, this coefficient has since been revised downwards, as being seven times lower, between 50 and 80. It is now measured at either 67 or 72.

THE PRESIDENT: Having two distinct values is hard to understand, especially with today's resources.

HUBBLE: This is due to two different calculation methods.

THE PRESIDENT: So, the two methods don't match, which is annoying. Which means that at least one of the two methods must be wrong. Would you like to add anything, Mr Noé?

NOÉ: A fundamental point about redshift. But before that, let me make two comments. The first is that not only, as you pointed out, do physicists vacillate between two values for the Hubble constant, but this so-called constant value is not constant. According to the ΛCDM model, i.e. the official cosmological model based on the theory of Relativity, this constant value would have changed over time to remain consistent with distant observations. Clearly, this so-called Hubble constant is flawed.

THE PRESIDENT: I'll let you take responsibility for what you say. And your second comment?

NOÉ: My second comment concerns the meaning of this law. It seems surprising that galaxies move away faster the further away they are. But in a classical explosion, this is perfectly normal. The higher the velocity of a particle, the further it travels, so there's nothing surprising about that. Similarly, **the faster a galaxy is ejected, the further it will travel. This is** reflected in the following formula[37]:

$$D = V.K \quad \text{with} \quad K = \frac{1}{H}$$

Where D is the distance, V the velocity, H the Hubble constant and K a time constant (about 15 billion years).

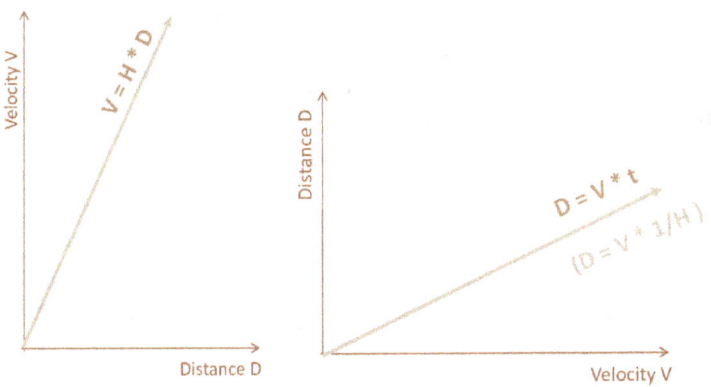

FIGURES 43: SPEED AS A FUNCTION OF DISTANCE OR DISTANCE AS A FUNCTION OF SPEED?

---

[37] "Gravity vs. Dark Energy, about the Expansion of the Universe", https://www.scirp.org/journal/PaperInformation.aspx?PaperID=81830

EINSTEIN: Except that with your formula, you believe yourself to be at the center of the Universe. In the relativistic ΛCDM model, there is no center to the Universe.

GALILEO: This was also Copernicus' position. He believed that the Universe was infinite, and that there could be no center.

NOÉ: Perhaps this is one of the reasons why he delayed publication of his work, fearing the wrath of the Church. But for Copernicus, the Universe was static. In the Big Bang context, the Universe is dynamic. In a classical explosion, there is obviously a center, the place where the explosion occurred. This is not incompatible with the fact that all particles are moving away from each other, not only those going in the opposite direction, but also those going in the same direction. Each particle thus sees all the others moving away. This certainly doesn't mean that each particle is at the center of the explosion. The same applies to galaxies: each one seems to be moving away from the others, but this in no way means that each one is at the center of the explosion. Unlike the relativistic ΛCDM model, in the neo-Newtonian model, there is indeed a center, but it's not possible this way to know where it is.

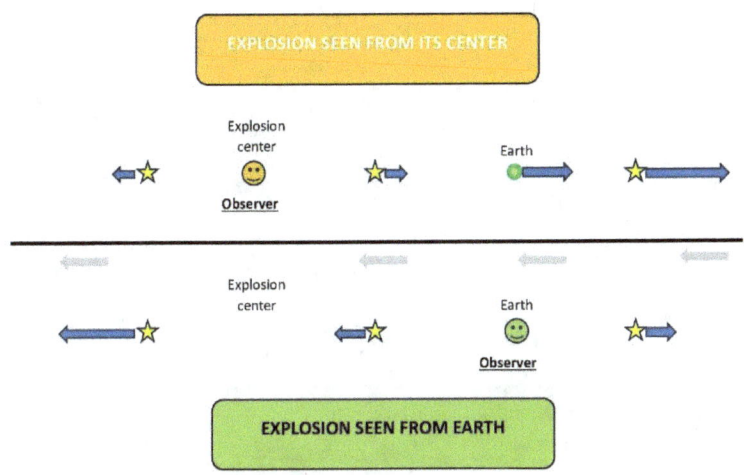

FIGURES 44: EXPLOSION SEEN FROM THE CENTER OF THE EXPLOSION AND FROM THE GROUND

THE PRESIDENT: We can't get you to agree. And what was your fundamental point.

NOÉ: That's going to take a little longer to explain.

THE PRESIDENT: So, let's take a short break before we listen to you.

GALILEO, WAKE UP!

29. RED SHIFT

THE PRESIDENT: Following on from the discussion on the Hubble constant, let's hear your fundamental point, Mr Noé.

NOÉ: This fundamental point concerns the calculation of redshift

THE PRESIDENT: Redshift is an increase in wavelength. It is caused by the movement of the light source away from the observer, by the Doppler effect or by the expansion of the universe. It is characterized by the variable " z ". We'd like to hear what you have to say about red shift, and ask you to be as concise as possible.

NOÉ: If I may, I'd like to correct that definition. Rather than an increase in wavelength, **it's an increase in period**.

THE PRESIDENT: Since one is exactly proportional to the other, it doesn't seem to make any difference.

NOÉ: Not exactly. If we consider the speed "c" as constant, it wouldn't change anything. But if this speed "c" of light is not precisely constant, that changes things. What the instruments measure is not a wavelength in meters, but a period in seconds, or rather the frequency, which is the number of alternations in a second. They don't measure a length in meters.

THE PRESIDENT: As you wish.

NOÉ: The other point to underline in this definition is that, according to the theory of Relativity, redshift could be generated by the expansion of the universe. In the relativistic sense, the expansion of the universe is not galaxies moving away from each other in an infinite space, but space itself expanding, intersidereal space creating intersidereal space.

EINSTEIN: You're thinking in terms of classical physics, where space exists outside and prior to all matter. What you haven't grasped is that it's matter that creates space-time. At the moment of the Big Bang, space-time didn't exist. It was this initial explosion that created space-time, and galaxies are moving away from each other as a result of this generation of space-time. The image of a raisin cake in the oven is a good illustration. The cake rises without the raisins rising. The raisins are the galaxies, the dough is space-time.

NOÉ: Comparison is not reason. In the cake in the oven, there's always the same amount of dough. If the cake puffs up, it's because it incorporates the surrounding air. There is no spontaneous generation of matter or dough. In your vision of the expansion of the universe, interstellar space initially doesn't exist, then is generated out of nothing, which is extremely bizarre. What's more, this spontaneous generation would only take place between galaxies, not between stars.

THE PRESIDENT: Please. Doesn't this take us away from the explanation you wanted to give about the red shift?

NOÉ: Indeed. In neo-Newtonian mechanics, redshift arises solely from the Doppler-Fizeau effect generated by the motion of galaxies in pre-existing space.

THE PRESIDENT: To be clear, according to Relativity, galaxies undergo the expansion of space and the red shift is not due to the Doppler effect. **And in neo-Newtonian mechanics, galaxies move through space and the red shift is due to the Doppler effect.**

NOÉ: That's right. Note that relativistic demonstrations of redshift usually go through this often hidden step: $\frac{c}{c-u} = 1$, with u representing the speed of the source relative to us, the observer. For this fraction to be true, the numerator and denominator must be equal, i.e. the velocity u is equal to zero, i.e. there is no movement of the source, so no redshift is possible. All these relativistic demonstrations of redshift are therefore invalid.

THE PRESIDENT: You're the one who says so. How could these demonstrations have been validated if they had been? But we'll examine your criticism of these demonstrations[38]. Are you yourself proposing another approach?

NOÉ: Yes, absolutely. **According to neo-Newtonian mechanics, the formula giving source velocity as a function of redshit is**[39] :

$$V = \frac{z}{1+z} c$$

with V the relative speed of the source with respect to the observer, z the redshift measured by the observer and c the speed of the photon with respect to the source.

The result is that, for the same redshift, the deduced velocity is lower than that calculated with the relativistic formula. And this is consistent with the velocity addition formula seen in Fizeau's experiment

THE PRESIDENT: Do you have any experience or observation to back you up?

NOÉ: Confirmation comes from the Pioneer 10 and 11 probes. You may have heard of these probes, which were sent into space with engraved gold plates intended to be deciphered by extraterrestrials. When the trajectory of these probes was no longer corrected by engines, an anomaly was detected: a stronger-than-expected deceleration. A deceleration, or braking acceleration, is a continuous decrease in speed. Using the relativistic redshift formula, the calculated value is different from the measured value. This difference is called the Pioneer anomaly. Using the neo-Newtonian redshift formula, the

---

[38] https://www.youtube.com/watch?v=BJyKmXiBUbk
[39] **"A New Formula of Redshift vs. Space Expansion and Dark Energy,"**
https://www.scirp.org/journal/paperinformation.aspx?paperid=107231

velocity and therefore the velocity variation of each probe corresponds exactly to what was measured[40].

THE PRESIDENT: In a way, the Pioneer anomaly is to neo-Newtonian mechanics what the Mercury anomaly was to Relativity. Let's not forget, however, that the cause of the Pioneer anomaly has now been officially given: it's said to be an internal flow of photons.

NOÉ: It took them ten years to come up with this photon explanation, in desperation if you will.

THE PRESIDENT: Anything to add?

NOÉ: Yes, one last paradox. According to the theory of Relativity, ultra-distant galaxies can move at superluminal speeds, i.e. faster than the speed of light. This contradicts the second principle of the theory of Relativity, the direct consequence of which is that nothing can travel faster than the speed of light. And when a theory contradicts itself, it's because it's incoherent, and therefore invalid.

EINSTEIN: It's not the galaxies that are moving at superluminal speeds in three-dimensional space, it's space-time itself that's growing. This space-time is expanding and, at the edge of the Universe, it's expanding faster and faster, until it exceeds the speed of light. It's not the galaxies that are moving, it's space-time that's expanding.

NOÉ: So, as seen from the inhabitants of these galaxies, we would be moving faster than the speed of light for them, that's extraordinary...

THE PRESIDENT: That's enough, let's stop this debate. Before we move on to the next expert, to get back to the original problem, what relationship did you make between the red shift and the Hubble constant?

---

[40] **"The Pioneer Anomaly explained by the Processing of the Doppler Effect"**, https://www.gsjournal.net/Science-Journals/Research%20Papers-Relativity%20Theory/Download/7330

NOÉ: The so-called Hubble constant for ultra-distant galaxies is calculated on the basis of an erroneous relativistic redshift formula. The distance-dependent velocities of galaxies would have to be recalculated on the basis of this new redshift formula.

THE PRESIDENT: What a program! Now let's hear from the next expert on an even more fascinating subject: black holes.

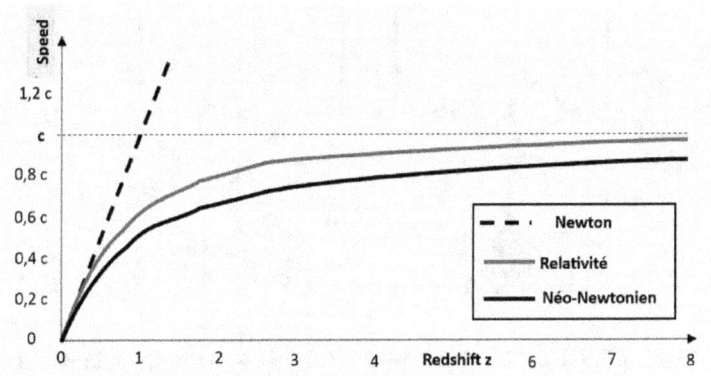

FIGURE 45: CALCULATION OF SPEED AS A FUNCTION OF REDSHIFT Z

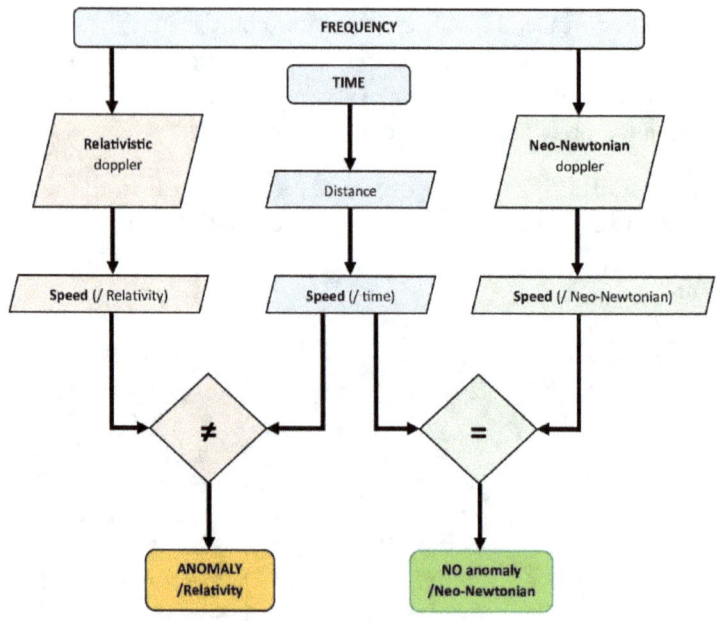

FIGURE 46: RELATIVISTIC AND NEO-NEWTONIAN REASONING

30. BLACK HOLES

THE PRESIDENT: We're now going to talk about the best-known and most enigmatic object - or rather concept - in the Universe: the black hole. By definition, a black hole emits no light, and is invisible to us.

THE REPORTER: Just a semantic clarification. Black holes are a relativistic term. The term was coined by a North American physicist more than fifty years after the development of the theory of General Relativity. They used to be called "occluded stars", "dark bodies", "dark", "black" or "invisible" stars.

THE PRESIDENT: Yes, this is not a new concept in the scientific world. Among the general public, the film Interstellar has done much to popularize it. It could be the gateway to other worlds, so-called white fountains, but so far, this is pure science fiction, purely hypothetical. We'll now hear from our expert, an Englishman, I believe.

THE REPORTER: He sends his apologies, as something came up at the last minute.

THE PRESIDENT: It's unfortunate, but nothing serious, I hope. What would you suggest?

THE REPORTER: Failing that, there would be a Frenchman.

THE PRESIDENT: Well, let's get started. Please introduce yourselves.

LAPLACE: Marquis Pierre Simon de Laplace, French mathematician, astronomer, physicist and politician since the Napoleonic era.

THE PRESIDENT. Very well, then. Please tell us about your contribution to our knowledge of black holes.

LAPLACE: Together with the Englishman John Michell, we independently hypothesized the existence of a star made up of so much matter that no photon of light could escape from it. As no photon of light could escape, it would be invisible to us.

THE PRESIDENT: If the concept of a black hole was speculative at the time, it nevertheless shows that it had been formulated long before Relativity. How would a black hole be proof of Relativity?

EINSTEIN: For many reasons. First of all, it's bounded by the event horizon, known as the Schwarzschild radius.

NOÉ: This radius corresponds to the black hole's boundary. Defining the size of the black hole is no proof of the theory of Relativity. In Newtonian mechanics, it is also possible to calculate the expression of this boundary from the centrifugal force.

EINSTEIN: Except that within this Schwarzschild radius, a black hole corresponds to a gravitational collapse of the space-time continuum.

NOÉ: Precisely, this gravitational collapse poses a problem: the curvature of space-time would become infinite at this singularity. But the physical world is not mathematical: there are no infinite elements. In Newtonian mechanics, there is no such thing as an infinite black hole.

EINSTEIN: The best proof of the existence of black holes is the deflection of light, with, for example, the aptly named Einstein Cross.

NOÉ: The existence of black holes has indeed now been proven, with the observation of Sagittarius A*, M87* or, as you say, Einstein's Cross as just one example. But their existence doesn't prove the theory of Relativity any more than it does Newtonian mechanics. The latter also explains these phenomena, along with the gravitational deviation of the photon, as we discussed in connection with Eddington's eclipse observation. On the other hand, since the deviation is twice as small according to Newtonian mechanics, this means that for the same observation of light deviation, either the black hole is twice as big, or the star is twice as far away as predicted by the theory of Relativity.

EINSTEIN: Another proof of Relativity is the gravitational redshift. Light passing close to a black hole will undergo a redshift due to the strong gravitational field. Observations confirm this redshift, and Newtonian mechanics cannot explain it.

NOÉ: Newtonian mechanics, maybe not, but in neo-Newtonian mechanics, the speed of the photon is not exactly constant, so it's normal that the photon can slow down under the effect of a very large gravitational force. On the other hand, the relativistic explanation of redshift with a constant velocity is more difficult to understand.

EINSTEIN: You're the one who doesn't understand it. Another piece of relativistic evidence is black hole mergers, which generate gravitational waves in space-time. And these gravitational waves have been observed.

THE PRESIDENT: An excellent transition to introduce gravitational waves, which will be explained by the next expert. Before that, do you have anything to add, Mr. Noé?

NOÉ: We've seen matter ejected in ultra-fast jets of plasma. But nothing should come out of a black hole, not even light.

EINSTEIN: These jets can, under certain circumstances, be emitted from the poles in a complex process.

NOÉ: Emitted or ejected from a black hole, that's indeed surprising.

EINSTEIN: And how would you interpret it?

NOÉ: In neo-Newtonian mechanics, one hypothesis - but it's **only one avenue to be explored - is that the black hole, which is just a black star, would be a star with a very high rotational speed, as quasars can be. As a result, the star would be flattened at its poles. These poles could then be below the Schwarzschild radius, allowing matter to be ejected.**

EINSTEIN: Pure speculation on your part, you have no basis in fact.

NOÉ: It's a pure hypothesis, I quite agree. But doesn't the theory of Relativity also make speculative projections such as the relativistic "evaporation" of black holes, i.e. their evanescent disappearance?

THE PRESIDENT: Sorry, gentlemen, we'll leave black holes there. What I take from this is that black holes can also be explained by Neo-Newtonian Mechanics and that they are not a clear proof of Relativity. On the other hand, let's be careful that the infinite singularity doesn't turn into a tomb for this theory.

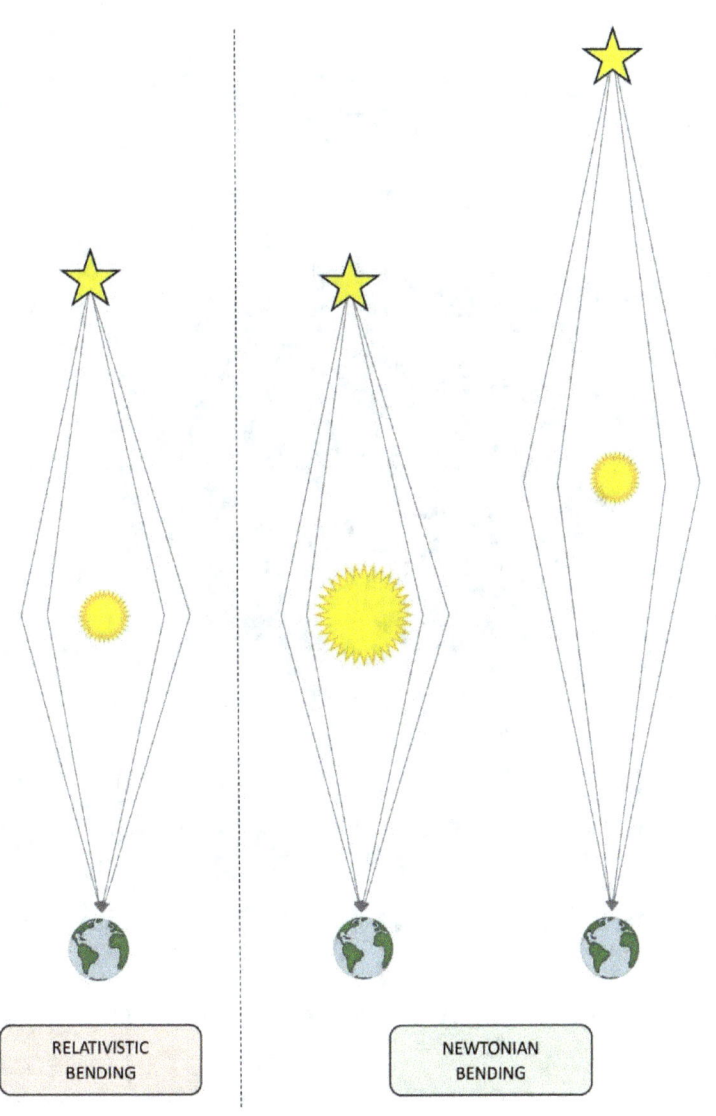

FIGURE 47: RELATIVIZED VS. NEO-NEWTONIAN DEVIATION

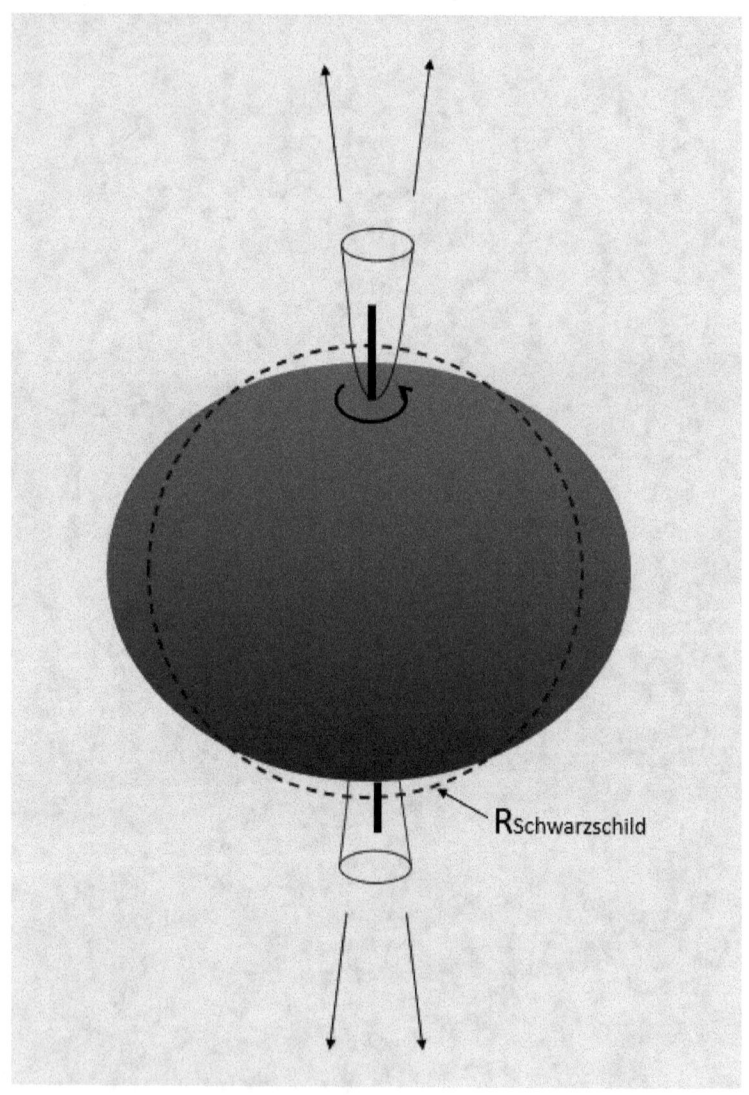

FIGURE 48: HYPOTHESIS OF A BLACK HOLE?

31. GRAVITATIONAL WAVES

THE PRESIDENT: Good morning, please introduce yourself.

WEBER: Good afternoon. My name is Joseph Weber, I'm co-inventor of the maser, the ancestor of the laser, and above all, I'm the first observer of gravitational waves.

THE PRESIDENT: Let's remember that, according to Relativity, a gravitational wave is an oscillation in space-time propagating at the speed of light from its point of formation. What does your experiment involve?

WEBER: I've designed and built several "resonance antennas". These are huge cylindrical bars fitted with sensors. The principle is quite simple: when a train of gravitational waves passes by, the bar deforms in the same way as space-time, and the sensors record these deformations.

THE PRESIDENT: What did you observe?

WEBER: On several occasions, deformations of the bar attest to the passage of gravitational waves.

THE PRESIDENT: That seems clear enough. Has your observation been reproduced by other independent teams?

WEBER: Yes, but they wouldn't confirm my results. Some even called me "crazy" in a public statement. If I had been, would I have been asked to provide one of my gravity detectors for an Apollo mission to the Moon?

THE PRESIDENT: We're not interested in personal attacks. More specifically, what were their criticisms of your observations?

WEBER: They wrongly accused me of seeing "signals in the noise".

THE PRESIDENT: I don't understand, could you clarify?

WEBER: These sensors are extremely sensitive to detect the passage of a gravitational wave through space-time. The variation in length of a bar cannot exceed one millionth of the size of an atom. So, thanks to the very high sensitivity of these sensors, the raw recording yields a series of disordered lines - in other words, noise. The key is to be able to extract the masked signal of the gravitational wave's passage from the noise, which they were unable to do.

EINSTEIN: Allow me to intervene. It's true that, according to Relativity, the vibration of space-time due to a gravitational wave is so infinitesimal that I myself thought it could never be detected. In fact, I've often wondered about their very existence. It's an extremely delicate experiment to perform.

NOÉ: Your questioning of their existence shows above all that gravitational waves are not fundamental, that they are merely a possible consequence of the theory of Relativity.

THE PRESIDENT: Thank you, Mr. Einstein, for your honesty. Please continue.

EINSTEIN: Following Mr. Weber's, shall we say, controversial results, other set-ups were developed, based on another principle inspired by Michelson's interferometer. These involved highly sensitive suspended mirrors separated by long steel tubes. And here, the results were not only conclusive, but repeated by independent teams.

NOÉ: What you don't say is that, as in Weber's observation, since the detectors are extremely sensitive in order to detect the passage of a wave, the recording also gives disordered deformations, i.e. noise. The criticism levelled at Weber, that he invented a signal in the noise, could just as easily be levelled at you. As all the teams use more or less the same computer program to extract a signal from noise, it's not surprising that from time to time they obtain an identical extracted signal.

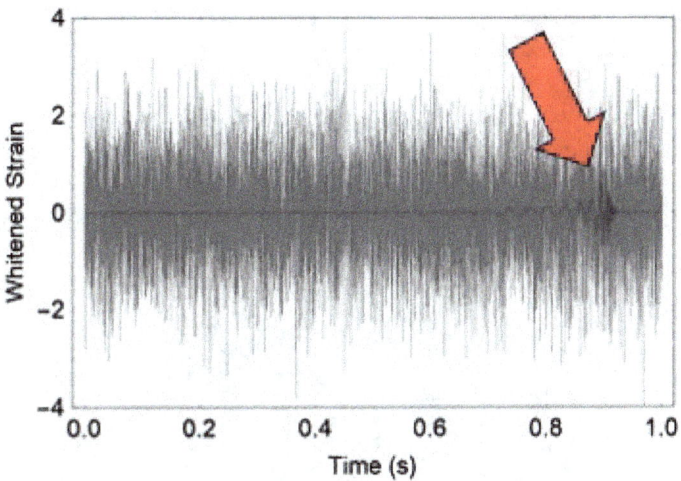

FIGURE 49: EXTRACTION (ACCORDING TO THE ARROW) OF A GRAVITATIONAL WAVE FROM DEFORMATION NOISE

EINSTEIN: I understand that it bothers you, but these teams obtained this identical signal from different noises, different initial data. It's clear that there was something in common, and that commonality is the signature of the passage of a gravitational wave.

NOÉ: A very brief passage, on the order of a tenth of a second.

EINSTEIN: Nonetheless, it's the signature of gravitational waves, which originate in particular from the merging of black holes.

NOÉ: It's unverifiable! It takes nearly a year of continuous observation to conclude that a black hole is present at a precise point in the Universe, and we manage to know in a tenth of a second that the gravitational wave comes from the merger of two unobservable black holes. Bravo!

EINSTEIN: It's obviously the theoretical framework that allows us to interpret the observed signal as the consequence of the merger of two black holes. But, depending on the form of the extracted signal, there may also be other cases, such as the merger of a black hole with a neutron star. In such cases, radiation and particles could be released.

NOÉ: All these representations in the media format our minds to make us believe that we're dealing with relativistic gravitational waves, just as, in other times, the cinema showed space travel where the passage of time slowed down. But these are only artist's images, only animations made by graphic artists.

FIGURE 50: TRUST IN ME!

I would have preferred a signal that moved a needle in an indisputable and not computer-processed way.

EINSTEIN: Maybe our computer scientists, engineers and technicians will get there one day.

THE PRESIDENT: You may dispute the interpretation, but you don't seem to be totally unwilling to admit that these signals are real.

NOÉ: Assuming the extraterrestrial origin of certain vibrations, couldn't they come from the flow of radiation emitted, for example, when celestial bodies merge? **This flow of radiation could be radio**

wave photons, capable of passing through a steel tube and striking the suspended mirrors. A gamma-ray burst is an extremely energetic explosion, usually lasting a few seconds, in a broad frequency band, particularly in radio frequency. The **energy of these radio-photons would be sufficient to set these oscillating mirrors in vibration**[41].

THE PRESIDENT: Your counter-proposal is rather hypothetical.

NOÉ: I agree, but it's more realistic than when the media broadcast the "song" of gravitational waves. There's no air in space, so there's no song transmitted.

KEPLER: Sounds don't exist in the divine heavens because of the slowness of movement on a human scale. The song of gravitational waves may be inaudible to the human ear, but harmonious to celestial beings. Do you know the quadrivium?

NOÉ: In ancient times, these were the four mathematical sciences: arithmetic, geometry, music and astronomy.

KEPLER: Yes, because they're linked. I believe in the harmony of the cosmic world and the harmony of music, which reflects that of the Universe and its Creator. For example, according to the mathematical characteristics of its elliptical orbit, do you know which harmony should be attributed to the Earth?

NOÉ: Unless I'm mistaken, mi...fa...mi...

KEPLER: That's right, fa for famine and mi for misery.

NOÉ: The misery you're talking about isn't indigence, but, I imagine, war and disease?

KEPLER: Such is the condition of Man on Earth.

---

[41] **"Gravitational waves or particle radiation?"**, https://www.physicsessays.org/browse-journal-2/product/1588-12-olivier-serret-gravitational-waves-or-particle-radiation.html

NOÉ: This relates more to Man than to Earth. On a planetary level, on Gaia's level if you like, the misery at the moment is more demographic pressure and ecological disasters.

THE PRESIDENT: That's a long way from our subject. To come back to gravitational waves and conclude, let's just be careful that their little music doesn't become a dangerous siren song to lead cosmic physics explorers astray. And let's hope that the next topic will be less about singing and computer animations and more about direct observation.

32. EXTRA-GALACTIC DARK MATTER

THE PRESIDENT: Good morning, please introduce yourself.

ZWICKY: Hello, Fritz Zwicky, astrophysicist and discoverer of Dark Matter.

THE PRESIDENT: That's why we've asked you here. First of all, why do you call it "dark" matter?

ZWICKY: I came up with the concept of "missing" matter. It wasn't until some twenty years later that some people renamed it dark matter, with the obvious risk of confusion with the black hole.

THE PRESIDENT: Indeed, the term "missing matter" is probably more explicit, but we'll keep the actual name here. Could you briefly define the characteristics of Dark Matter?

ZWICKY: Dark Matter, or Missing Matter, is a hypothetical form of matter, invisible, non-interacting with matter, stable, inhomogeneously distributed. Its main characteristic is that it is massive, attracting galaxies gravitationally.

GALILEO: It reminds me of ether. It's a speculative substance, invisible, impalpable, immobile. The difference is that it is uniformly distributed. Its main characteristic is to explain the propagation of light waves.

ZWICKY: Except that with the photon, dear Sir, we no longer need the hypothesis of an etheric propagation medium to explain light propagation. But we still need Dark Matter to explain the movement of galaxies.

THE PRESIDENT: Mr. Zwicky, could you tell us about your comment?

ZWICKY: I studied the motion of seven galaxies within the Coma cluster of Berenice's Hair, and noticed that the speed of the galaxies was significantly higher than would have been expected. The cluster of galaxies had to be extremely massive to prevent certain galaxies from moving away from the group; it had to be four hundred times more massive.

GALILEO: 400 times is unimaginable! The work of a mental pygmy, no doubt.

ZWICKY: Who the hell is this spherical nard? It's the difference in value that's been calculated.

THE PRESIDENT: Let's have some restraint, gentlemen! 400 times more massive, you say! It's not even a difference, it's got nothing to do with it! It's incredible! How do you explain it?

ZWICKY: For this purpose, I had formulated the hypothesis that the cluster was in dynamic equilibrium. The thermal agitation that pulls the galaxies apart had to be compensated for by the gravitational attraction that draws them together.

THE PRESIDENT: In the context of the continuous expansion of the Universe, we might question the relevance of this hypothesis of dynamic equilibrium over time, but go on.

ZWICKY: At the time of my work, the expansion of the Universe was still only a vague hypothesis, and the very term Big Bang would not be "invented" or popularized by English astronomer Fred Hoyle until some fifteen years later, and then only in derision. At the time, the stability of the universe prevailed.

THE PRESIDENT: You were right to recall the context of the time.

ZWICKY: Then, it's true that I underestimated the luminosity of the galaxies, seeing the cluster as less massive than it is. Also, relying on Edwin Hubble's work on galaxies, which saw galaxies moving away seven times faster than they actually do, I also overestimated the distance of the galaxies and therefore their speed.

THE PRESIDENT: I understand that, since the galaxies observed are actually much closer, the observed speeds are much lower than you had estimated.

NOÉ: If I may say so, the gravitational mass of this cluster of galaxies is now estimated to be greater, much greater, at least ten times greater. Distances and velocities are estimated to be smaller. It is from the multiplication of these major estimation errors that the hypothesis of missing mass has emerged. I insist, the Dark Matter hypothesis comes from an estimate that is 400 times wrong.

THE PRESIDENT: Don't gloat too much, the Dark Matter hypothesis has not been disavowed.

NOÉ: That's debatable. Since then, not only has the visible mass been significantly increased, but the mass of the hot gases has been added, which could not be seen in Mr. Zwicky's day. **With the rectified visible mass and the mass of the hot gases, there's no longer any missing mass, so there's no need to invoke Dark Matter.**

THE PRESIDENT: You said it yourself. The Dark Matter hypothesis, even if not firmly established, remains valid. And why is that, Mr. Zwicky?

ZWICKY: While the mass thought to be missing has indeed been found in the Coma cluster, the same cannot be said of most other observed clusters. It is now estimated that Dark Matter represents six to seven times the visible matter.

THE PRESIDENT: How was it estimated this time?

ZWICKY: As my friend Françoise said, with the discovery of the cosmological background, you need exotic dark matter to be able to form dark galaxies, i.e. galaxies that don't emit light. In this context, we calculated that ordinary matter could not exceed 5% of the total mass of the Universe.

THE PRESIDENT: It's a total paradigm shift. Initially, Dark Matter was to be used to explain the stability of galaxy clusters. Now, it's being used to explain the initial formation of the Universe, with invisible dark galaxies made of dark matter. Quite a hypothetical model, don't you think?

NOÉ: A model that's 400 times wrong!

GALILLE: Fi! The work of a cuistre!

THE PRESIDENT: Please! Why keep this model, Monsieur Zwicky?

ZWICKY: There's another reason for maintaining the Dark Matter hypothesis: the rotation curve of galaxies.

THE PRESIDENT: This is an excellent transition to the next hearing, which, for once, will be given by an expert. Thank you for your participation.

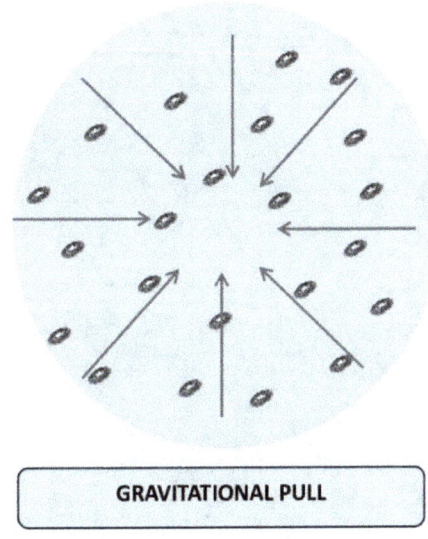

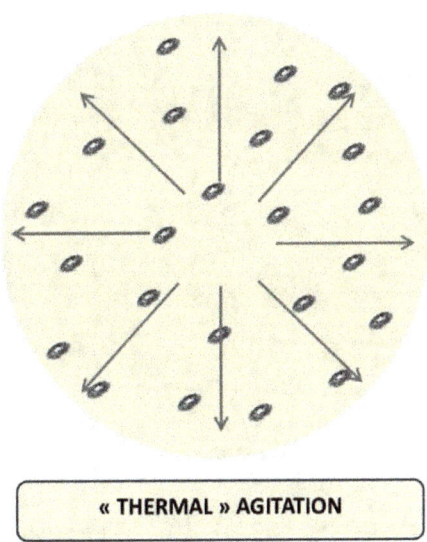

FIGURE 51: GRAVITATIONAL ATTRACTION AND "THERMAL" AGITATION

GALILEO, WAKE UP!

33. INTRA-GALACTIC DARK MATTER

THE PRESIDENT: Good morning, dear lady, please introduce yourself.

RUBIN: Good morning Madam President, good morning Gentlemen. My name is Vera Rubin, and I'm an astronomer. I've studied galaxies in all their aspects: their own motion and in clusters, their distribution, their stability, their rotation speed, their merging. As you can imagine, I'm an expert on galaxies.

THE PRESIDENT: Could all this work have earned you a Nobel Prize?

RUBIN: Yes, some people have said that. Personally, I'm shocked by the low number of recognized women in science, and I've spent my life fighting against certain scientific institutions. That said, my results are more important to me than my reputation. After all, the important thing is to find something new.

THE PRESIDENT: Could you tell us about your discovery concerning the rotation curve of galaxies?

RUBIN: Simply put, a galaxy is a vast collection of stars, gas and dust revolving around a center. The most common galaxy, the classic spiral, takes the form of a disk with a bulge at its center. The rotation speed of the stars in the galaxy changes according to their distance from the galaxy's center. This is known as the rotation speed curve of the stars in the galaxy, and is abbreviated to the galaxy's rotation curve. Observations often show something surprising: the rotational velocity of stars and gas remains constant and sometimes even increases at distances further from the center.

THE PRESIDENT: Why is that surprising?

RUBIN: In a rigid plate rotating around its center point, or on a classic merry-go-round, the further you are from the center, the faster you turn.

THE PRESIDENT: A galaxy is not a rigid whole.

All the stars in a galaxy revolve around the center at roughly the same speed. As a result, the stars closest to the center make several complete circles, while the distant stars only make one complete circle.

GALILEO: This was the vision of Pythagoras, for whom all the planets move at the same speed, in harmony with each other.

THE PRESIDENT: Why is that so surprising?

RUBIN: Because, according to Newtonian mechanics, this speed should decrease as you move away from the center. In Newtonian mechanics, the more distant circles rotate more slowly. To give an example, let's take a look at our solar system, with the Sun at its center. In this system, Mars, a planet further from the Sun than the Earth, rotates more slowly around the Sun. Mercury, the closest planet to the Sun, orbits faster than the Earth.

GALILEO: In my day, there was a saying about distant planets that travel a longer distance and move more slowly: "He who wishes to travel far spares his mount".

RUBIN: Newtonian mechanics explained it by a lesser force of attraction at greater distances. Applying this model of planetary rotation (around the sun) to stars (around the bulge of the galaxy), the near-constant speed of rotation of these stars, whatever their distance from the center of the galaxy, cannot be explained by visible matter alone. This suggests the existence of invisible matter distributed throughout the galaxy.

THE PRESIDENT: And what would this invisible matter be?

RUBIN: Since it doesn't emit light, we call it Dark Matter.

THE PRESIDENT: The same one Professor Zwicky imagined to describe the movement of galaxies in a cluster?

RUBIN: That's right.

THE PRESIDENT: So Dark Matter would be between galaxies in clusters, in dark galaxies at the beginning of the universe, and in galaxies for the rotation curve. It's everywhere, but we can't see it anywhere, we can't bring it to light, it's infuriating in the end. What do you think, Monsieur Noé?

NOÉ: One fact is that the rotational velocity curve of galaxies is globally constant. But the comparison with planets around the Sun is inappropriate. In fact, in the solar system, the mass of the planets is nothing compared to the mass of the Sun; practically all the mass of the system is concentrated in the Sun. It is therefore normal, according to Newtonian mechanics, for planetary rotation speeds to decrease with distance from the sun.

THE PRESIDENT: Why shouldn't what's normal for the orbit of planets in the solar system be normal for the orbit of stars in a galaxy?

NOÉ: In a galaxy, the mass of its bulge represents only a fraction of the galaxy's total mass. **The mass of the stars that revolve around the bulge contributes significantly to the mass of the galaxy. We have therefore proposed the following model for the mass distribution of galaxies**[42] :

> **The mass distribution of the galaxy's disk, excluding the bulge, could be a $1/r$ function,**
>
> with r the star's distance from the center of the galaxy

It's because of this mass distribution that the curve is constant in a galaxy.

---

[42] **"The flat rotation curve of our galaxy explained within Newtonian mechanics"**, https://www.ingentaconnect.com/content/pe/pe/2015/00000028/00000002/art00007

THE PRESIDENT: Why is the weight divided into 1 on r, where does that come from?

NOÉ: Imagine a jet of matter that begins to spin. Over **time, it would take on the shape of a disk, thicker in the center, thinner around the edges**[43].

RUBIN: Except that, according to the latest models, the distribution is of exponential "-r" form.

NOÉ: Galaxy models have come a long way, if only to progressively take into account non-luminous or faint matter such as black holes, dying stars, brown dwarfs, asteroids and comets, dust, interstellar gas, neutrinos and so on. We have also seen that, for the galaxies of the Berenice Hair, the first estimates were largely wrong about their mass. By a factor of 400. But there's not much difference between an exponential model and a 1 in r model. Perhaps one day the official model will come closer...

GALILEO: As far as the concept of Dark Matter is concerned, there are a few things that bother me. On the one hand, there are as yet no experiments to prove the existence and properties of this Dark Matter. But, as you know, experience is essential in science, it's fundamental. What's more, many different things can be explained by this Dark Matter. We agree that the same cause produces the same effects. But different effects come a priori from different causes. The equilibrium of galaxy clusters, the formation of the universe and the rotation curve of galaxies - all very different effects - are all explained by the same cause, in this case dark matter. Each of these three effects could have its own cause, and the explanations could be different.

THE PRESIDENT: The multiplicity of causes is not necessary, but it does give food for thought, so we'll leave it at that for now. For the rest of the hearings, we're continuing on the dark side of cosmology. The next one deals with dark energy.

---

[43] **"Rotation curve of our Galaxy,**

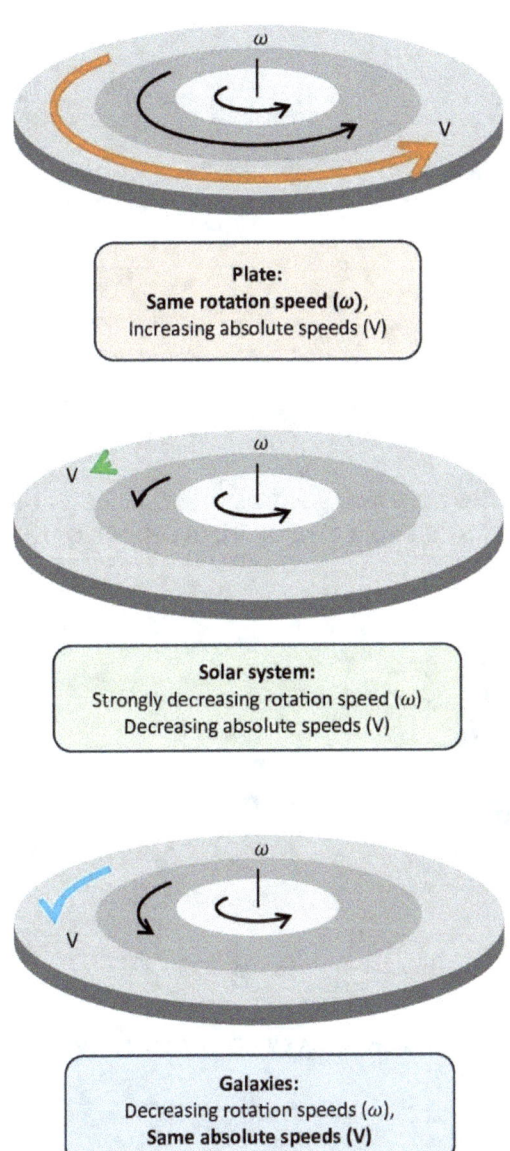

FIGURES 52: 3 TYPES OF ROTATION SPEED

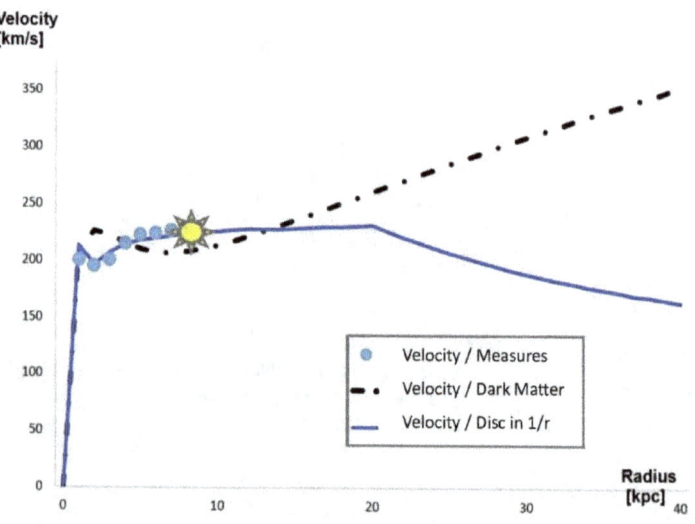

FIGURE 53: ROTATIONAL SPEEDS OF STARS AS A FUNCTION OF THEIR DISTANCE FROM THE CENTER OF THE GALAXY

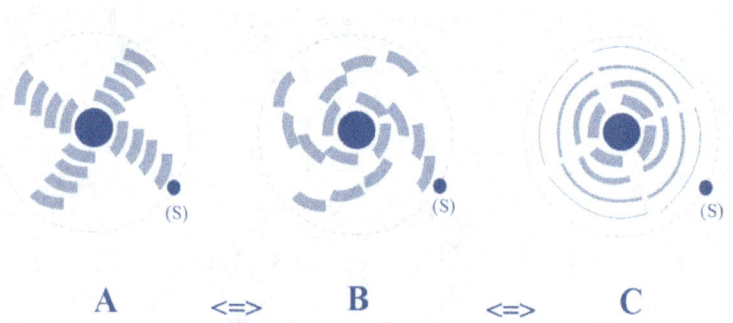

FIGURES 54: 1/R GALAXY FORMATION HYPOTHESIS

34. DARK ENERGY

THE PRESIDENT: Please bring in the next expert.

THE REPORTER: The expert we were expecting couldn't make it, held up in the other world. His work is widely recognized, and Saul was awarded the Nobel Prize in 2011 " for the discovery of the accelerating expansion of the Universe through observations of distant supernovae". Supernovae are stars that all explode with the same intensity. This is why they are used as standard candles, in other words, as beacons or standards of luminous distance.

KEPLER: I observed one in 1605.

THE PRESIDENT: That's right, we've been calling it the Kepler Supernova ever since it was the last one to be observed in our Galaxy. It's a pity Saul can't come, as the subject is complex. So we'll take it one step at a time. Could you start by explaining the concept of dark energy?

THE REPORTER: Dark energy is a hypothetical form of energy uniformly filling the Universe.

THE PRESIDENT: A hypothetical form, like dark matter, right?

NEWTON: Or like ether

THE REPORTER: Indeed, its existence has not yet been established by experiment. It is assumed that it behaves like a repulsive gravitational force, i.e. it doesn't attract bodies, but repels them.

THE PRESIDENT: I'm not sure the audience understands it properly, could you explain it differently?

THE REPORTER: Yes, of course. You throw a ball into the air, it falls back down under the effect of gravity. Well, let's suppose that the ball doesn't fall back down and continues to rise, or that it falls back down very gently, in either case, as if attracted by a force aloft. To say it's attracted by a force aloft is the same as saying it's repelled by a force on the ground, which pushes the ball upwards. A kind of antigravity.

NEWTON: Then it would be a bit like electrical forces, which can either attract bodies or repel them. The idea is interesting, but negative gravity has never been observed in our solar system.

THE PRESIDENT: In our terrestrial environment, no, not in our galaxy either, but it would be in deep space. Anyway, to get back to our subject, how does dark energy confirm Relativity?

EINSTEIN: Believing the Universe to be static, i.e. initially opposing the idea of an expanding Universe, I had imagined adding a cosmological constant Λ to my equations. This constant would guarantee stability by opposing the effects of gravity, i.e. it would oppose any contraction of the universe. This was the most obvious mistake of my life, as a static universe was really too special a case. I eliminated this calculation artifice in the light of the observations made at the time, attesting to the expansion of the universe.

THE PRESIDENT: Let me ask you again: how does dark energy confirm Relativity?

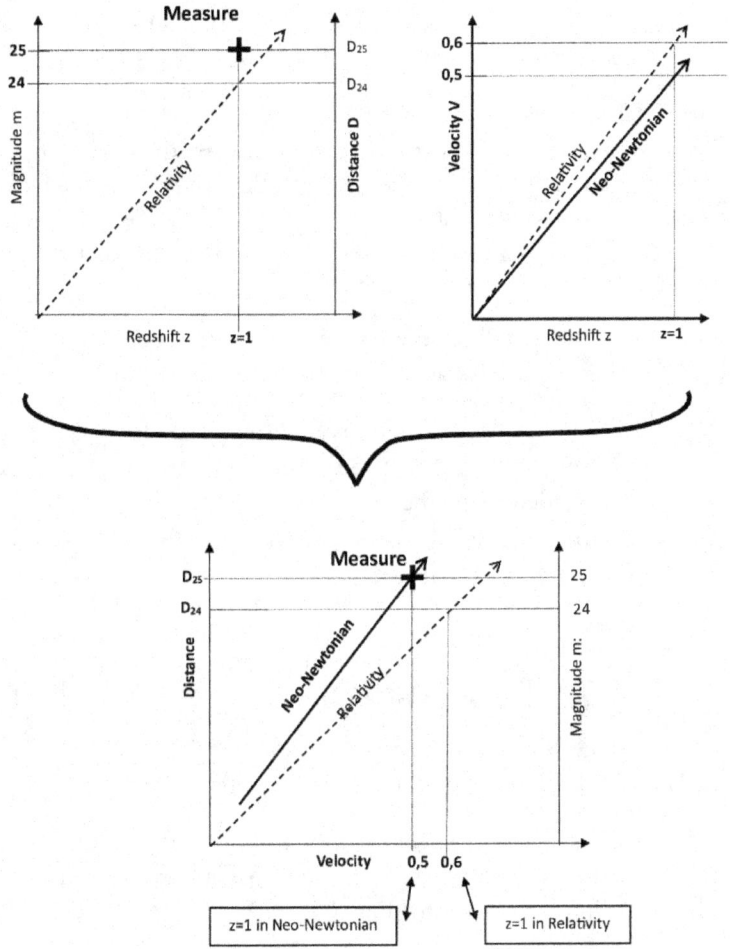

FIGURES 55: DETERMINING DISTANCE AS A FUNCTION OF REDSHIFT

EINSTEIN: I'm coming to that. Finally, this constant is reintroduced into the equations of the most common cosmological model, the ΛCDM model, because it represents vacuum energy, which is the equivalent of dark energy.

NEWTON, *articulating*: The e-nergy of the vacuum... isn't that antinomic? A spontaneous generation of energy from nothingness is difficult to understand and contradicts the conservation of energy.

EINSTEIN: Indeed, it may seem surprising at first. But the role of the cosmological constant Λ can be explained by the evolution and reinterpretation of the role to be given to this constant in relativistic equations. Initially, its function was to counteract gravity to obtain a static model. Today, this vacuum energy, or dark energy, helps explain the Universe's increasing expansion.

THE PRESIDENT: We're already into interpretation, so let's get back to the facts. What was observed?

THE REPORTER: So we've observed distant Type Ia supernovae.
- On the one hand, from their apparent luminosity, we were able to determine their distance.
- On the other hand, by using redshift, we were able to determine their speed of separation

THE PRESIDENT: And what was the conclusion?

THE REPORTER: At low light levels, the redshift is lower than expected.

THE PRESIDENT: And how do you interpret this?

THE REPORTER: If you find more redshift in the past this would imply that the expansion has been slowing down and hence there is more mass in the universe to slow it down.

THE PRESIDENT: Apparently, you're explaining the opposite of what was observed. Can't you comment on it in the right order?

THE REPORTER: Of course, that's just what Saul said. To put it in the right order, as you're asking me to do: if you see less redshift than expected from the current rate of expansion, that implies that expansion has been slower in the past and has increased. In other words, the rate of expansion is increasing. Hence the need for dark energy expansion forces.

NEWTON: This is a very complex subject. When you refer to the redshift predicted from the current expansion rate, this is also based on Hubble's law and its current value, is that correct?

THE REPORTER: Absolutely!

NEWTON: If there's an acceleration of expansion, it's because the distances covered increase with time, so the inverse of the Hubble constant increases with time, i.e. the Hubble constant decreases with time.

THE REPORTER: Yes, without a doubt!

NEWTON: If the Hubble constant decreases over time, it's because the Hubble constant used to be larger.

THE REPORTER: Well, certainly!

NEWTON: And if in the past Hubble's constant was greater than it is now, according to Hubble's law, the speed of expansion was greater than it is now.

THE REPORTER: Um..., obviously!

NEWTON: And if the speed of expansion was greater than it is now, then the speed of expansion is decreasing, which is contradictory to the hypothesis of an accelerating expansion.

THE PRESIDENT: Top, top, top! Let's stop here, please! There's obviously a fallacy somewhere, on one side or the other. We'll have to take a fresh look at this reasoning. And what about you, Mr Noé?

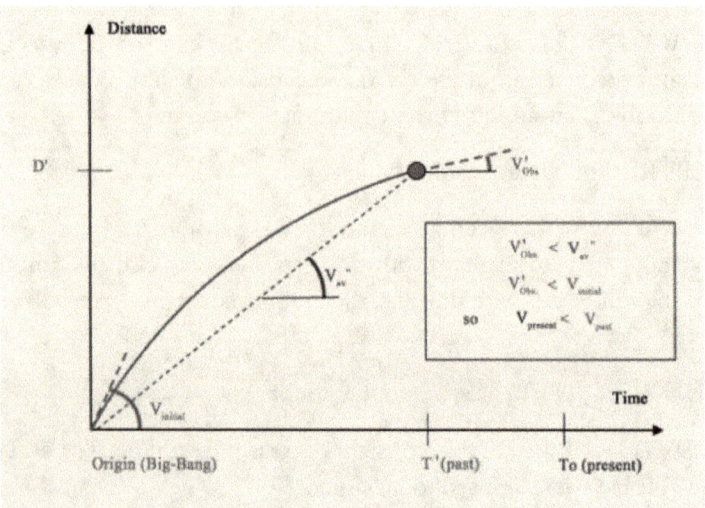

FIGURE 56: GALAXY DISTANCES AS A FUNCTION OF TIME

NOÉ: I agree with Newton, who seems to have raised a paradox. And **when there's a paradox, it's because one of the hypotheses is wrong, in this case the relativistic redshift rule.**

THE PRESIDENT: What do you mean? Can you elaborate?

NOÉ: According to neo-Newtonian mechanics, for the same redshift reading, the object's velocity is lower than that predicted by the theory of Relativity. In the case of distant Supernovae, this means that their instantaneous velocity is lower than their average expansion velocity defined by galaxy distance. In other words, **it means that the speed of expansion is gradually decreasing**[44].

THE PRESIDENT: With your theory, you arrive at the opposite conclusion to that of Relativity.

NOÉ: That's right. But beyond the conclusion, to get there, I didn't need to do any semantic pirouetting or bring in any hypothetical dark energy. If the rate of expansion decreases over time, this is simply

---

[44] **"Gravity vs. Dark Energy, about the Expansion of the Universe",**
https://www.scirp.org/journal/PaperInformation.aspx?PaperID=81830

due to the effect of gravity, which gradually slows down the explosive effects of the Big Bang.

THE PRESIDENT: Presented like this, your model seems simpler than the ΛCDM model, where we have to add dark energy and dark matter, so that we would only be able to grasp 5% of the total mass-energy of our Universe. Nevertheless, the relativistic interpretation of accelerated expansion has been crowned with a Nobel Prize. And there's one final point that could work in favor of the relativistic interpretation: the cosmic microwave background, which we shall now examine.

GALILEO, WAKE UP!

35. THE COSMOLOGICAL MICROWAVE BACKGROUND

THE PRESIDENT: Good morning, you're the last expert we're going to listen to. Please introduce yourself.

PENZIAS: Arno Penzias, engineer and doctor of physics. With Robert Wilson, we accidentally discovered the Cosmic Microwave Background.

THE PRESIDENT: What exactly did you observe?

PENZIAS: We have observed electromagnetic microwave radiation coming from space in a way that is considered uniform and isotropic. Simply put, we observed a constant signal of microwaves coming from all directions in the sky.

THE PRESIDENT: How was that interpreted?

PENZIAS: This microwave signal corresponds to an emission source from a body at a temperature of 2.7 Kelvin, extremely close to absolute 0 Kelvin. This would mean that this very cold source could be found at the very edge of the Universe.

THE PRESIDENT: How does this confirm the theory of Relativity?

PENZIAS: The Cosmological Background is considered to be one of the strongest pieces of evidence in favor of the Big Bang and, by extension, the theory of General Relativity.

NOÉ: You can reject the theory of Relativity and still accept the Big Bang hypothesis.

PENZIAS: According to commonly accepted hypotheses, shortly after the Big Bang, the recombination process released enormous quantities of light. According to the theory of Relativity, the expansion of the Universe caused the wavelength of existing photons to be stretched as a result of the stretching of space-time. It is this original light that is now being detected in the form of microwaves. It has an extremely high redshift.

GALILEO: Religion may have made a mistake when it tried to enter the realm of science. In the opposite direction, doesn't Science make the same mistake when it wants to describe the Creation of the Universe?

NOÉ: To say that the cosmic microwave background comes from the expansion of space-time is an interpretation. How did it come to be said that this micro radiation was originally visible light?

PENZIAS: Well, by spectroscopy, we were able to determine that this radiation had a redshift z of about 1090. This shows that it was originally a luminous radiation.

EINSTEIN: What we're seeing are photons of light emitted over 14 billion years ago, and which have therefore taken 14 billion years to reach us. That's why they seem to form a "background". But we'd better call it a "cosmological horizon", because according to Relativity, the Universe has no center and no boundaries. Hence this roughly uniform, isotropic diffusion.

NOÉ: And yet, if you look at the observed image in detail, there's not isotropy, but anisotropy, meaning that there are properties that change depending on the direction observed. There is a red drift in one half of the sky and a blue drift in the other. What is generally communicated to the general public are corrected images, where the effect of motion has been subtracted (see illustration). If we take the raw data, we can clearly see that we are moving in relation to this diffuse background.

EINSTEIN: But it's not incompatible. The fluctuations correspond to variations in the density of the early universe, as predicted by models based on General Relativity and cosmic inflation.

NOÉ: No, it's not these fluctuations I was talking about, but **the temperature dipole, showing a global orientation. Our galactic system is moving at a speed of 570 km/s in relation to this Diffuse Cosmological Background**[45].

EINSTEIN: It's the Doppler effect caused by our Galaxy's course through space-time.

NOÉ: It's interesting that you're now taking up the Doppler effect. And if everything were relative, there would be no reference space. In which case, why would the solar system and the Galaxy move 570 km/s relative to this Cosmological Background? This is not consistent with the principle of relativity, where there is no privileged frame of reference.

EINSTEIN: You're the one who's being incoherent. How would you interpret it?

NOÉ: In the same way as you, namely that our Galaxy moves at a speed of 570 km/s in relation to this diffuse background. But unlike the theory of Relativity, in Neo-Newtonian Mechanics there is indeed a center of the Universe and a privileged frame of reference, so there's no inconsistency.

THE PRESIDENT: I note a point of agreement. It could be the last word.

NOÉ: There's a lot more to be said about ultra-distance phenomena. For example, some time ago, the James Webb telescope observed ancient barred galaxies at very high redshift that closely resembled today's galaxies. In line with the ΛCDM model, the self-expansion of the universe and the predictions of the theory of Relativity, these

---

[45] **"A New Formula of Redshift vs. Space Expansion and Dark Energy,"**
https://www.scirp.org/journal/paperinformation.aspx?paperid=107231

barred galaxies were estimated to be some ten billion light-years away at the time of their luminous emission. In other words, they would be extremely young, on the order of a few billion years. The problem is that this requires us to revise our model of galaxy formation. According to this model, these ultra-distant galaxies are supposed to be smaller and less luminous.

EINSTEIN: But the reasoning is simple. The redshift z gives the speed, which gives the distance, which gives the age.

THE PRESIDENT: What other explanation would you have, Mr. Noé?

NOÉ: There are a number of possible explanations. However, it's the link between redshift and speed that's wrong. **According to neo-Newtonian redshift calculations, these high redshift galaxies are only 3 to 4 billion light-years away. They would therefore not be ultra-distant, and therefore not so young. There's no need to fundamentally revise the galaxy formation model. The question** is, what's wrong with the James Webb Telescope's observation of galaxies: the galaxy formation model, or the theory of Relativity?

THE PRESIDENT: Well, gentlemen, you won't agree. Save your arguments and your eloquence for tomorrow, when each of you will make a case for your theory. It's been another busy day. I wish you a pleasant evening and look forward to seeing you in the morning.

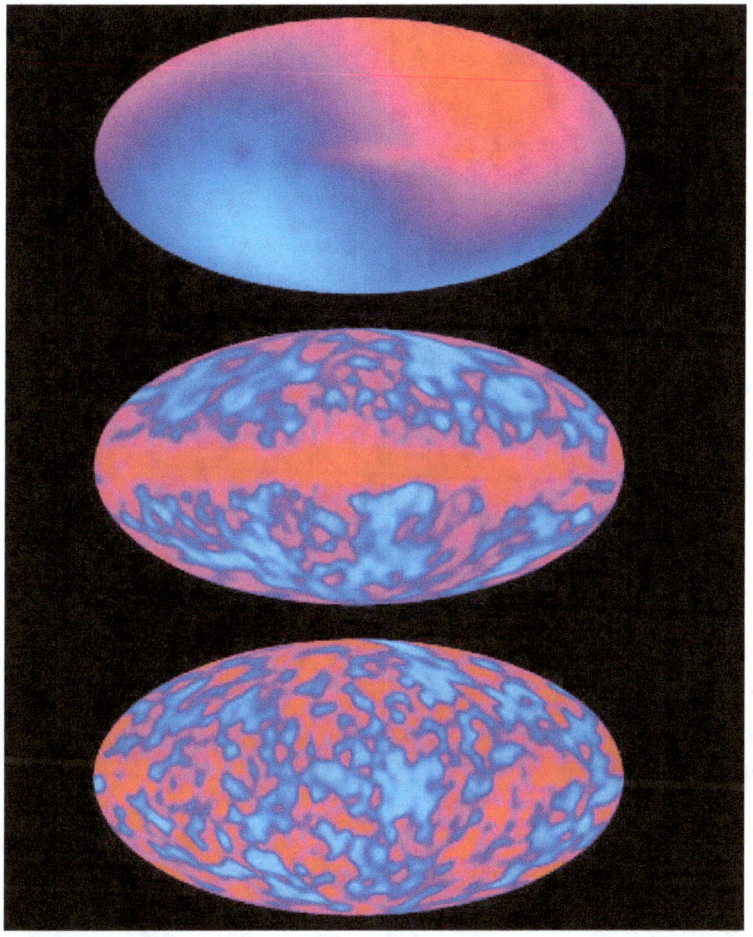

FIGURES 57: RAW AND REPROCESSED IMAGES OF THE COSMIC MICROWAVE BACKGROUND

Top: raw results of COBE satellite observations. Directional movement is observed.
In the middle, the effect of the dipole has been removed. The effect of motion has been subtracted.
Below, the effects of the dipole and Milky Way eliminated. This is the image generally broadcast to the public, suggesting that there is no preferred direction (in line with the theory of Relativity).

GALILEO, WAKE UP!

DAY FOUR

CONCLUSIONS

GALILEO, WAKE UP!

36. THE RELATIVIST PLEA

THE PRESIDENT: Good morning to you all. I'd like to thank you for your attendance and for your diligence in listening to the various experts who have come before you. With this fourth and final day, we have come to the end of our discussions, which have been rich in learning and questioning. We must now conclude, leaving each of the two protagonists to synthesize and defend their own theory before you for the last time. The drawing of lots has designated the theory of Relativity to be defended first. It is also, if we need reminding, the theory that has been established for centuries. We're listening.

KEPLER (*aside*): Since he's universally recognized as a genius, Einstein is well assured. But if you don't doubt, you can't be sure of anything.

GALILEO (*answering in a low voice*): Yes, doubt is the father of creation. Let's listen to it.

EINSTEIN: Madam Chairman, members of the Evaluation Committee,

Today, I stand before you to plead once again for one of the most revolutionary scientific achievements in history: the theory of relativity, which I initiated and which has been enriched by thousands of researchers, including many Nobel Prize winners. This theory, which has profoundly transformed our understanding of the universe, rests on solid foundations and has been confirmed time and again by rigorous experiments and observations. It was developed in the face of the intrinsic limitations of Newtonian mechanics and its obvious shortcomings. For example:

- The Newtonian rule of addition of relativistic speeds, which can lead to a result greater than the speed limit of light and which is invalidated by Fizeau's experiment with moving water.

- The Newtonian rule of adding up the speeds of light invalidated in Michelson's experiment.

- The Newtonian determination of the precession of orbits, which lacks the precision to determine that of Mercury

Allow me then to remind you of the arguments that support the indisputable relevance of the theory of relativity.

First of all, special relativity, formulated in 1905, is based on two fundamental postulates: the Galilean principle of relativity and the principle of the constancy of the speed of light. We won't go into detail here on the first postulate, which is recognized by everyone, including the opposing party. The second postulate, asserting that the speed of light is the same for all observers regardless of their motion, has been repeatedly confirmed by experiments, including those involving subatomic particles accelerated to speeds close to that of light. These observations clearly show that the speed of light is a universal constant, confirming the very essence of special relativity.

Here are a few proofs that validate the accuracy of this theory:

- Time dilation: Experiments like the one with the clocks in an airplane and the one with muons confirm the time dilation predicted by special relativity.

- Contraction of lengths: measurements of the distances covered by velocity muons and those of Michelson's experiment show a contraction of lengths in the direction of motion, in accordance with special relativity.

- Equivalence of mass and energy: The famous equation $E=mc^2$, which expresses the equivalence of mass and energy, has been confirmed by nuclear reactions and particle physics experiments.

Then, general relativity, developed in 1915, provided a profound new understanding of gravity. This theory proposes that gravity results from the curvature of space-time caused by the presence of mass and energy.

The predictions of general relativity have been confirmed by precise observations, such as

- The bending of light around massive objects, as is the case when observing eclipses or gravitational lenses.

- The precession of Mercury's perihelion, which curiously no expert came to testify at these hearings, was confirmed by precise observations.

What's more, the theory of relativity has stood the test of time and technological advances.

- The gravitational redshift observed in light from massive objects, already seen from the Sun, has been confirmed around black holes.

- The accuracy of our global positioning system, GPS, depends on the relativistic corrections required to take account of gravitational effects and space-time deformations.

- Finally, observations of gravitational waves, first detected in 2015, are in perfect agreement with the predictions of general relativity concerning the violent motions of massive objects such as black hole mergers.

In conclusion, the theory of relativity is not based on theoretical postulates alone, but has been underpinned by decades, if not centuries, of research, observation and experimental validation. The theory revolutionized our understanding of space, time and gravity, and continues to guide modern scientific research. It is recognized as such by the scientific community. As members of this commission, I urge you to confirm the depth and indisputable validity of the theory of Relativity, which remains - and for a long time to come, no doubt - one of the fundamental pillars of modern physics.

Thank you very much.

THE PRESIDENT: We'd like to thank you for your brilliant presentation.

37. THE NEO-NEWTONIAN PLEA

THE PRESIDENT: As I was saying, as fate would have it, we're going to conclude these three days of expert hearings with a neo-Newtonian twist. As with the defense, this gives you the advantage of having the last word. We're listening.

NEWTON (*aside*): Whatever he argues, I think all bets are off, that his position will remain solitary and inaudible to the scientific community, which can't turn its back like that.

KEPLER (*answering him in a low voice*): I much prefer the virulent criticism of a single man of mind to the thoughtless approval of the masses.

GALILEO (*in a low voice*): The authority of a single competent man, who gives good reasons and certain proofs, is better than the unanimous consent of those who know nothing about it. Let's hear him out.

NOÉ: Thank you, Madam Chairman and members of the committee,

I stand before you today to begin by recalling the major advances made by Newtonian mechanics in science, a theory that has been a pillar of classical physics for centuries. It's already important to recognize the substantial achievements and enduring areas of applicability of Newtonian mechanics, which is not yet to be consigned to the dustbin of history - far from it, given its many current uses.

Allow me to highlight a few points in favour of Newtonian mechanics:

> - Precision within the limits of everyday life: Newtonian mechanics remains extremely precise in situations where velocities are well below the speed of light and gravitational fields are moderate. In other words, Newtonian mechanics is used for almost all practical applications, from the modelling of everyday movements to calculations concerning the solar system and sending men to the Moon.
>
> - Simplicity and clarity: Newtonian mechanics offers a simple, intuitive conceptual framework for understanding the motion of objects. Its fundamental laws, such as the law of universal gravitation and the three laws of motion, are easy to teach and understand, making it a powerful educational tool. Distances are fixed, time flows uniformly.
>
> - Compatibility with Quantum Mechanics: Newtonian mechanics is compatible with Quantum Mechanics, which deals with the behaviour and properties of subatomic particles. There's a continuity of explanation from the microscopic to the macroscopic, because from a physics point of view, there can't be two antagonistic approaches to the world we live in.

In short, Newtonian mechanics has stood the test of time and continues to be an invaluable tool used on a daily basis and for almost all modelling in science and engineering.

Admittedly, it has a few marginal weaknesses, but these have been exploited by the theory of Relativity. The latter explored a new field, that of ultra-velocities close to the speed limit of light. But this theory is based on revolutionary and unintuitive concepts, such as the constant speed of the photon and its corollary, variable time. We won't dwell on the weakness of the evidence provided by the various experts during these days, as there are so many misinterpreted experiments: Michelson's incomplete experiment, Sagnac's invalidation of the constancy of the speed of light, cosmic distances varying according to the mode of measurement, Langevin's imaginary predictions, the false results of the round-the-world flight, Eddington's selection of observed data, resynchronized GPS without appeal to Relativity, the uncertainty of the results of Fizeau's experiment, the interpretation of the variable half-life of muons, the error in demonstrating the redshift formula, the infinite and unphysical singularity of black holes, the very brief noise of gravitational waves extracted from ambient noise, dark matter everywhere but never found, the addition of equally hypothetical and distant dark energy, the evaluation of the distance of ultra-distant galaxies observed by the James Web telescope, and so on. Other important problems such as velocity additions, ovoid orbits and the measurement of pulsar dispersion could be discussed frankly over a drink. In short, this theory would never have been faulted, but it is so incoherent that, by dint of additions and shaky scaffolding, it itself admits that it can only explain 5% of the physical world, with 95% remaining unknown, unobservable and hypothetical. It is also regrettable that the relativist plea did not take up or respond to the various criticisms voiced during the conference, as if it had been written before all these experts, whom we thank here, were heard.

Moreover, confidence in this theory of Relativity is beginning to crumble in the so-called scientific community, and some researchers say they are prepared to examine an alternative theory if it is presented as comprehensive. It would be superhuman, not to say impossible, to propose an alternative theory on one's own, dealing in a detailed and indisputable way with all areas of physics and astrophysics.

However, compared with other alternative theories, neo-Newtonian mechanics has the advantage of taking all the contributions of Newtonian mechanics, used in the vast majority of cases, and extending them to the case of ultra-high speeds, relying solely on the difference, both qualitative and quantitative, between gravitational mass and inert mass. Although neo-Newtonian mechanics is, as it stands, potentially acceptable, it lacks indisputable proof. These could include experiments such as those presented at the conference:

- **Michelson's experiment as perceived by an observer moving in relation to it**

- **The Sagnac experiment with a modified route that's easier to interpret**

- **Fizeau's experiment with a rotating silicon disk instead of translating water**

- **The chromatic effect or not of Shapiro delay**

- **Exploiting and observing the refraction of the sun's corona, thanks to satellites like Gaia**

- **Sending another Pionner-type satellite to measure its distance from the object.**

- **The ovoid trajectory of the star S2 and certain comets**

It is these cosmic experiments and observations, dear members of the commission, dear experts and dear readers, that we suggest you carry out, or have carried out, in order to form an informed opinion on the proposals and perspectives of this neo-Newtonian mechanics.

Thank you for your attention.

38. EPILOGUE

THE PRESIDENT: Thank you for those two speeches, which sum up the arguments put forward over the last three days. I would like to thank the audience for their presence and the interest they have shown, even if on one occasion it was a little noisy. I'd also like to thank all the experts who so clearly shared their experiences and observations with us. Of course, I'd also like to thank the members of this committee, who shared their immense knowledge and incisive analysis with us. Everyone has been able to give their own interpretation of the same physical phenomena. Now it's up to you to recall them and draw your own conclusions. For our part, we at IPPC are giving ourselves time to reflect.

GALILEO, *in himself*: Ah, I'm not angry that these interminable debates are over, they were wearing me out in the end. For a while, my rheumatism was forgotten, but now it's back with a vengeance, making me suffer with every movement. I can't help thinking of the writings of dear Plato. Didn't he define science as true opinion with reason? Science would then remain, in other words, no more than a more or less justified belief. This wise man knew his world. If nothing else, this shows that the truths enunciated by the ancients have survived the centuries. And there's **only one way** out of ill-founded belief and opinion, **to reach scientific reason, true science: experiment, experiment and more experiment!** And in a room, in a laboratory, where you can vary the parameters you control. Not outdoors, where observations can always be interpreted according to our respective beliefs, as these debates have shown once again, if proof were needed. And the way to interpret the results of experiments is to transcribe them into mathematical form. Because, no, mathematics is not an invention of the devil - quite the contrary. The laws of the Universe are written in mathematical language. It enables us to decipher the living letters of the book of Nature, the work of God. Will my disciples believe me when I tell them what I've just experienced? And that smell of camphor that invades me again. Finally, I feel I'm returning to Italy, to Tuscany, to my home. But I also feel I'm leaving...

Galileo Galilei died on Wednesday January 8, 1642 at the venerable age of 77.

POSTFACE

Dear reader, did this extraordinary story really happen? Is it even possible? Certainly, it's possible if you still believe in variable time and time travel. If not, if you have doubts about these beliefs, if you have shared some of the criticisms levelled at the giant with feet of clay that is the theory of Relativity, if you still rely on the main concepts of Newtonian mechanics to describe and predict the physical phenomena that surround us, then this booklet will not have been written in vain.

This fictional appetizer is intended to whet the appetite of scientific readers, so that they can sample some of the demonstrations provided in the footnotes. These are articles on Neo-Newtonian Mechanics that have already appeared in online journals, some peer-reviewed, some not. Even if they are written in English, I hope he'll appreciate and share this 'new' cuisine, which aims to bring Newtonian mechanics back up to date.

GALILEO, WAKE UP!

ACKNOWLEDGEMENTS

My warmest thanks once again to **Brigitte G-D** *for her assiduous proof-reading and comments on this fiction.*

TABLE OF ILLUSTRATIONS

Figure 1: Falling bodies 13
Figures 2: Ptolemy, Tycho-Brahe and Copernicus reference frames 20
Figures 3: Galilean reference frames 25
Figure 4: Force of gravity & Force of inertia 30
Figure 5: Tides 30
Figure 6: Galileo's Principle of Relativity 35
Figures 7: Second postulate of Relativity 35
Figures 8: Perceived light velocities 47
Figure 9: Measurement of light deflection by the sun 53
Figure 10: Measurement processing diagrams 54
Figures 11: Shapiro effect 58
Figures 12: Actual (solid line) and expected (dotted line) results 64
Figure 13: Relativistic reasoning 64
Figure 14: Paths perceived from the ground and from space 65
Figure 15: Paths traveled from the ground and from the disk 70
Figures 16: Current and proposed configurations 71
Figure 17: Galaxy distances in Relativity theory 76
Figure 18: Galaxy distance in Neo-Newtonian mechanics 77
Figure 19: World tour results table 86
Figure 20: Relativistic and neo-Newtonian measurement processing 90
Figure 21: Eötvös coefficient 96
Figure 22: Free fall of a satellite 96
Figure 23: Relativistic and neo-Newtonian definitions of the Gamma factor 99
Figure 24: Bertozzi's results 99
Figure 25: Fringe displacement as a function of water velocity 105
Figure 26: Proposed experiment with a glass disk 106
Figure 27: GPS satellite resynchronization 110
Figure 28: Refraction of a disc as a function of wavelength 124
Figure 29: constant speed, variable distance and time 125
Figures 30: Gravitational field & Inertial field 128
Figure 31: Gamma factor as a function of speed 133

Figures 32: The two types of speed addition 136
Figure 33: Velocity additions according to the theory of Relativity 137
Figures 34: Speed additions according to Neo-Newtonian Mechanics .. 139
Figure 35: Ellipse vs. ovoid 145
Figure 36: Elliptical eccentricity of comets as a function of their perihelion 145
Figure 37: Measuring Pulsar Dispersion 150
Figure 38: Hypothesis on photon displacement? 153
Figure 39: Attractive gravitational mass and attracted inert mass 157
Figure 40: Synchrotron radiation 157
Figure 41: Convention on cosmological distances 162
Figure 42: Initial (500) and current (67 or 72?) Hubble constant values 164
Figures 43: Speed as a function of distance OR distance as a function of speed? 167
Figures 44: Explosion seen from the center of the explosion and from the ground 169
Figure 45: Calculation of speed as a function of redshift z 175
Figure 46: Relativistic and neo-Newtonian reasoning 176
Figure 47: Relativized vs. Neo-Newtonian deviation 181
Figure 48: Hypothesis of a black hole? 182
Figure 49: Extraction (according to the arrow) of a gravitational wave from deformation noise 185
Figure 50: Trust in me! 186
Figure 51: Gravitational attraction and "thermal" agitation 193
Figures 52: 3 types of rotation speed 199
Figure 53: rotational speeds of stars as a function of their distance from the center of the galaxy 200
Figures 54: 1/R galaxy formation hypothesis 200
Figures 55: Determining distance as a function of redshift 203
Figure 56: Galaxy distances as a function of time 206
Figures 57: Raw and reprocessed images of the Cosmic Microwave Background 213

Table of contents

PREFACE 3

PRELIMINARIES (GALILEAN) 7

   1. FALLING BODIES 9

   2. THE GALILEAN PRINCIPLE AND TRANSFORMATION 15

   3. THE LAW OF INERTIA AND GALILEAN REFERENCE FRAMES 21

   4. INERTIAL MASS AND GRAVITATIONAL MASS 27

   5. THE CONSTANT SPEED OF LIGHT 31

DAY ONE (RELATIVITY) 37

   6. THE THEORY OF SPECIAL AND GENERAL RELATIVITY 39

   7. NEO-NEWTONIAN MECHANICS 43

   8. LIGHT DEFLECTION 49

   9. THE SHAPIRO EFFECT 55

   10. THE MICHELSON EXPERIMENT 59

   11. THE SAGNAC EXPERIENCE 67

   12. VARIABLE DISTANCES 73

   13. THOUGHT EXPERIMENTS 79

   14. VARIABLE TIME 83

   15. THE MUON EXPERIMENT 87

DAY TWO (NEO-NEWTONIAN) 91

   16. THE MICROSCOPE EXPERIMENT 93

   17. BERTOZZI'S EXPERIENCE 97

   18. THE FIZEAU EXPERIMENT 101

   19. THE GPS 107

   20. INTERMEDIATE 111

21. 100 AUTHORS AGAINST EINSTEIN	117
22. ASYMPTOTIC SPEED S	131
23. THE TWO SPEED ADDITIONS	135
24. ELLIPSE *vs.* OVOID	141
25. THE PHOTON	147
26. ENERGY	151
27. THE STRENGTH	155
DAY THREE (COSMOLOGICAL)	159
28. HUBBLE'S CONSTANT (?)	161
29. RED SHIFT	171
30. BLACK HOLES	177
31. GRAVITATIONAL WAVES	183
32. EXTRA-GALACTIC DARK MATTER	189
33. INTRA-GALACTIC DARK MATTER	195
34. DARK ENERGY	201
35. THE COSMOLOGICAL MICROWAVE BACKGROUND	209
DAY FOUR	215
36. THE RELATIVIST PLEA	217
37. THE NEO-NEWTONIAN PLEA	221
38. EPILOGUE	225
POSTFACE	227
ACKNOWLEDGEMENTS	229
TABLE OF ILLUSTRATIONS	230

GALILEO, WAKE UP!

ISBN : **9798322172826**

Copyright © Olivier Serret, 2024